The Rise of the Scientist-Bureaucrat

Jose Luis Perez Velazquez

The Rise
of the Scientist-Bureaucrat

Survival Guide for Researchers in the 21st
Century

 Springer

Jose Luis Perez Velazquez
The Ronin Institute
New York, USA

ISBN 978-3-030-12328-4   ISBN 978-3-030-12326-0   (eBook)
https://doi.org/10.1007/978-3-030-12326-0

Library of Congress Control Number: 2019931522

This Springer imprint is published by the registered company Springer Nature Switzerland AG
The registered company address is: Gewerbestrasse 11, 6330 Cham, Switzerland

Old age is like climbing a mountain. The higher you get, the more tired and breathless you become, but your views become more extensive.

Ingmar Bergman

Preface

> *When a hypothesis enters a scientist's mind, he checks it by calculation and experiment, that is, by the mimicry and the pantomime of truth. Its plausibility infects others, and the hypothesis is accepted as the true explanation for the given phenomenon, until someone finds its faults. I believe the whole of science consists of such exiled or retired ideas: and yet at one time each of them boasted high rank; now only a name or a pension is left*
>
> Vladimir Nabokov, Ultima Thule

The words of Nabokov expose a succinct summary of the scientific work. Science, while continues to have the same vision as in antiquity namely the understanding of natural phenomena, has changed substantially from what once was. It is the practice of science, the performance of scientific research and its academic environment that has been somewhat transformed due to current socio-economic circumstances. As an aspect of society, science and its practitioners are not beyond the realm of contemporary transformations taken place in society driven by numerous causes, of economic origin in the main.

What will be described in the present narrative will no doubt surprise many readers, especially those not directly connected to the world of scientific research. Current times have witnessed the emergence of a monstrous machinery constructed around science and academia —machinery of bureaucratic or administrative nature most of it— which, as it is becoming apparent, does not increase the quality of research nor makes it easier for the scholar to navigate around this academic realm to do what should be done: research to understand natural phenomena. It is always reasonable to perceive and receive things as they come without being too judge-mental, for it is true that things evolve and one must adapt to the new situations, and the changes do not need to be qualified as better or worse —as we tend to do every minute of our lives, always judging and rating whatever comes to our senses. But, to be honest, I must admit that I have a reason, the only reason, why I would claim that the current situation in research is not as good as before; it is because scientists are not doing what they like to do, what they should be doing —research— rather, as it will become clear throughout this narrative, they spend much more time busy in administration and other chores. Nobody would complain if some bureaucracy occupies a little of our time, but the problem that I, and many others, see today is that these activities occupy most of our time as scientists, and we have little left for what we really like: to think, reflect on questions, perform experiments, analyse data and interpret results.

Some things that will be exposed may perturb or even upset some. Yet, all is true. All the specific events that appear in this volume really happened, and the only alterations have possibly been due to the fact that they are narrated after entering my subjective experience —we should never forget that books are personal objects— and as a result, they are presented after passing through a veil of comedy, as I have tried in many instances to emphasise their comic, many times absurd, nature. Because, even though these themes here treated are serious matter as they could direct many young people to turn one way or another, one can never forget the words that someone once pronounced: "do not take life too seriously because you will not get out of it alive". Nonetheless, I have tried to provide a fair view of the realm of the scientific enterprise in current times. As well, my purpose is not to criticise even though many things commented upon will sound like fierce criticism. But those comments are nothing but the description of the current status quo in academia and research. Nevertheless, and to be fair, throughout this volume we will explore some solutions to those apparent criticisms —in each section, there is a Possible Solutions subsection, where specific counsel for particular obstacles that researchers will find throughout their careers is given. Obviously, these recommendations are well known by professionals in the field, so it will not tell them anything new, but keep in mind this text is aimed at beginners and lay audiences who may not know these tricks we all have to perform to survive in academia.

It is not the purpose of this chronicle to distress readers or to discourage young students to proceed with a career in science. Not at all. In fact, it is just the opposite: these words have to be taken for what they are, a warning of what one will find in the execution of scientific research. Better to be informed than not informed or misinformed. The fulfilment of the vision of science —the understanding of natural phenomena— needs enthusiastic young students that, because of their keen interest in understanding nature, cannot do any other thing but to enrol themselves into the research population. We have to consider that, to some extent, the scientist and the artist have a crucial feature in common: the intense motivation. The artist expresses himself/herself through the artwork, the scientist via the investigation of problems and questions one thinks are important and should be resolved and found. Thus, the real scientist and the real artist are doomed to practice their trades, regardless of how difficult the situation may be, for such is their motivation. For them, neither art nor research is really a job. Naturally, there are others, not as motivated, that enter science more as though it were a job, for in the end it is a job: one is paid —although not much if you are in academia— and performs a service to society. Yet, those aforementioned extremely driven scientists will probably fleetingly hesitate the answer when asked what their job is, because to them research is not a job but rather a —how should we put it— sort of a hobby, what one does because the pleasure and satisfaction it brings are unsurpassed by other activities. It is for these reasons that one purpose of the book is the encouragement of the young to enter scientific research, despite what some sections of the volume may reveal. And here lies another purpose, perchance a hope: to spread the awareness such that the new recruits will realise the many times strange, paradoxical situations in the scientific endeavours and therefore will be brave enough to change it to what scientific research really should be, to what the scientist should do: research, experiments, reflect on problems. Because, as surprising as it may sound, these are things scientists do not perform in this day and age. Let's see why.

Oviedo, Spain   Jose Luis Perez Velazquez

Contents

Chapter 1
The Rise of the New Scientist—The Scientist-Bureaucrat

To start the story in the most straightforward manner that reveals the general facet of the current scientific enterprise, let us look at one typical photograph of a scientist that appears in the media, normally after the individual has won an award or something important has occurred to the scholar. What do you see? The researcher, probably in her/his laboratory, wearing a white coat and managing some laboratory utensils, perhaps a pipette, or looking through a microscope, or sitting next to some scientific equipment that, as it happens, she/he almost never uses any more and probably has even forgotten how to use them. For this scientist is an important one, a head of a laboratory —a P.I., for principal investigator, as they are called in scientific parlance— or perhaps the director of a department, an institute… In any event, someone who now leads a laboratory that houses a group of people. Yet, those research materials you see in the photograph around her/him are being used frequently —do not worry about it, they are not wasted— but by the trainees and technicians in the laboratory, not by the P.I.

But how can this be? This P.I., being in charge of the research in the laboratory, must do a lot of experiments, has to teach students and postdoctoral fellows — postdocs for short, both students and postdocs are considered trainees, this being a more general term— how to operate the sophisticated laboratory equipment, how to perform specific measurements, experiments, all those things the trainees hope to learn from the advisor so that when they leave the laboratory they can, one day, become PIs too. Well, they learn, but normally from a technician, or from a senior postdoc or research associate. The advisor, after becoming a P.I., has basically stopped doing experiments or analysing data. And here one encounters the first of the many paradoxes that will be found in this volume: while she (for the sake of convenience, let's assume the scientist photographed was a woman) was taught during the training period —for there was a time when the P.I. was a trainee too — how to perform experiments, how to conduct research very well, now the individual finds that she cannot do what she learned to do so well. This is a fact that young fellows who are thinking about going into science with the hope of doing research all their lives must realise; in the current times, and the reasons will be explored in what follows, the

© Springer Nature Switzerland AG 2019
J. L. Perez Velazquez, *The Rise of the Scientist-Bureaucrat*,
https://doi.org/10.1007/978-3-030-12326-0_1

scientist stops being a real researcher —that is, she who does experiments and evaluates data and all that — and turns into a scientist-bureaucrat after becoming enthroned to a laboratory. Now her time has to be spent in a myriad of chores that will impede her from doing what she was taught how to do so well —experiments, data analysis, in a word: research. Principal among the chores, is the search for money.

Hence a new type of research has started for her, not at the laboratory bench but in her office: the search for funds that will be used to pay salaries to technicians and trainees, to buy equipment, reagents, and all the many things an experimental laboratory needs. Writing grants has now become her major activity. Ask any independent scientist what the most time consuming of his/her occupations is, and the immense majority will answer in unison: grant writing. And it is not only the effort to write many grants, but what comes afterwards. Because once you write enough grants and obtain enough money, it means you have more funds to distribute, more administrative forms to fill out, more research and financial progress reports to submit to the funding agencies... Never underestimate the tremendous waste of time and dissipation of energy involved in the submission of grant applications and the reporting once you obtain the funds. In short, more chores that divert your attention from your real job, research in the laboratory. In the interviews in "Young, talented and fed-up" (https://www.nature.com/news/young-talented-and-fed-up-scientists-tell-their-stories-1.20872) you will see the common grievance mentioned many times in this book, the lack of time they have left to do research. Furthermore, in "Battling the bureaucracy hydra" [1] —the references are at the end of each chapter—you will learn of the ventures of Jörgen Johansson who, after being awarded a grant by the European Research Council, experienced the immense and unyielding monstrosity created around the research business; in his story you will learn that "the contract negotiations had to start immediately. With 19 different documents amounting to more than 150 pages to read or fill out, the contract process seemed like a many-headed bureaucratic hydra. I had to convert the text and budget of my application into a legal document", and other agonies he had to suffer in order to enjoy the awarded funds.

It is not only the search for funding and its management that becomes a preoccupation of a P.I. Among other affairs that keep her busy one finds the administrative tasks in the department, in the university or institution where the laboratory is located or affiliated. There is a myriad of administrative duties academics must perform these days. For instance, normally scholars are subjected to annual internal or external reviews so that the institution can follow in detail the performance of the individual. Many of these reviews involve the writing of substantial texts, which has triggered fear in the hearts of several scholars so much so that some have declined to go for something which they were entitled. For example, I have heard of some academics that refused to apply for promotion due to the immense package that one must fill out. My package to be promoted to Associate Professor —as an illustration of the phenomenon— consisted of 55 pages, and for that of my promotion to Full Professor I had to amass 121 pages (curiously, almost exactly double number of pages than the other, perchance denoting the double (?) importance of the title... even though my salary did not double after my promotion). Then we also have many other forms to complete, including experimental protocols for ethics approval, progress and financial reports to granting agencies, protocols for the use of certain controlled reagents, our own peer-review assessments of our colleagues in the institution, documents for disclosures of inventions and ideas... This short and of course incomplete list will give you an idea. As well, one more task to add is the teaching that universities impose once one becomes a Professor, normally starting as an assistant professor; but whether assistant, associate or full, that matters little, for the teaching load is, very commonly, considerable. Graduate and undergraduate teaching consumes hours not only of lecturing but also preparing the lectures, the syllabus, the exams, etc.

But her new occupations do not end with teaching, money and administration. There is more, much more. Now, as an independent scientist, she will be asked to judge others. After all, isn't she supposed to be an expert on a specific topic? Therefore, she is able to review papers and grant applications from scholars in the field. She will be asked to join review panels in funding agencies, and will be sent dozens of projects, grant applications that she will have to read, understand, and score. And due to her expertise in a specific area, she will be asked to review papers submitted for publication. She may be offered to be part of the editorial board of a journal too, which will impose on her more responsibilities such as handling papers submitted to the journal, finding reviewers for them, reading the reviewers' criticisms and comments (in reality when it comes to peer review these two words are synonymous), and coming up with a decision to reject or accept the paper for publication. This is called peer review, in the official jargon. A section later on (Chap. 5) will be devoted to this subject, as it is a fundamental aspect in the current scientific enterprise that needs to be improved. In short, scholars today spend much more time doing work for administration than doing laboratory work.

But, was she taught how to do all these administrative matters as a trainee? Especially, writing grants, and, once the funds are obtained, allocating the moneys, making sure there will be enough livelihood for the duration of the project. Some advocate that the trainees should be prepared to be immersed in the highly bureaucratic milieu of science —see for example the advice in "Why aspiring academics should do less science" [2]. I have seen trainees who were somewhat discouraged by the prospect of becoming an administrator. They entered science because they enjoyed doing research, and this is what they expected to do to the end of their days, but they find that academia is turning into an industrial enterprise. Just look at one recent advertisement for a Professorial post and inspect the requirements for this, apparently, academic job, emphasising income generation and other lucrative aspects: "Applicants must show evidence of experience and success in research/*commercial income generation*, engagement with the research *impact agenda, engagement with key stakeholders* in the Computational Neuroscience community, *and management* and *leadership* ability". What picture thus emerges in your mind after reading all these requirements in the advertisement? Doesn't it look more like an administrative type of job, rather than one that is supposed to discover causes and reasons for certain natural phenomena?

The growing bureaucratic invasion of academia is reaching a perturbing state of affairs that is motivating a large number of academics to warn about the perils of the administrative scene in science. There are abundant blogs, editorials and journal articles on the bureaucratic takeover of academia, for instance, "The irresistible rise of academic bureaucracy" published in The Guardian provides numbers of administrators infiltrating the academic world. Organizations are articulating their thoughts too. These are the words you can find in The Euroscientist: "Bureaucracy is spreading like the plague. It has now pervaded every aspect of scientists' lives; often to the point of choking the hardiest of investigators" (https://www.euroscientist.com/hacking-bureaucracy/). After these words, some advice is offered: "Yet, technology has evolved so much so that it now offers simple solutions to cut through the paperwork and make the scientific process more efficient, more collaborative and altogether smarter. Now, the time is right to take advantage of the opportunities afforded by technology and raise to the challenge of removing the hindrances brought by bureaucracy." This theme of the bureaucratic incursion into science and academia in general is the central theme that binds together all aspects that will be touched in this text, where we shall see how science is indeed turning into an industrial enterprise. Can this bureaucratic, corporative future be avoided? The Possible Solutions subsection below presents some advice, at least for the individual scholar to circumvent some administrative burden. As to more specific (or sometimes general) suggestions about reducing bureaucracy in the realm of science and academia, those pieces of advice will be offered in the different sections that will follow, as some are more specific and related to topics discussed in those sections.

Nonetheless, reduction of administration in science is a pressing task because bureaucracy is limiting the normal course of research. Two events that exemplify the role of administration in curtailing research are offered here. The first event occurred when we were about to start a collaboration with another institution on a project aimed at predicting the outcome of patients after brain injury. The colleagues from the other institute came to our place for a couple of days, tried the method, and apparently it worked well in a small set of patients (they brought some of their patient data), hence we decided to proceed further and to try it in many patients to assess the efficiency of the system we had devised; our colleagues were to go back to their institution and collect data from many patients. For this project to succeed, we needed to give them our software, computer programmes we coded to implement the technique, and besides they should be giving us some patient data. While our plans were to start immediately, as we were all very enthusiastic about the prospect, here at this point bureaucracy intruded. Because there would be exchange of codes and data, both institutions needed to sign some forms that go under various names, one of them is data sharing agreement. No research can start without such an agreement signed by all parties. It took between 6 and 12 months to have the two institutions sign this agreement, do not ask me why it took so long.

By the time it was finally signed, either our colleagues were not interested anymore or were too busy with other things, or had moved somewhere else (I do not remember these specific details), thus in the end nothing happened. Here we see how bureaucracy effectively stopped a project— or more properly did not allow it to develop, for it had not even started except for some very preliminary observations we had gathered during those two days —that could have been important for the treatment of brain injuries. Let me give you another, perhaps more trivial, example. Once we had some extra funds that we had to use within a limited time or we would lose them, therefore we decided to use the money to buy a piece of equipment we needed in the laboratory. We found the vendor, who sent us the quotation exposing the price, and we submitted the purchase form to our administration to buy it. But of course, nothing is simple when it comes to bureaucracy: we were told that since the equipment was expensive we needed to give them not one but two or three quotations. Mind you, we wanted this specific machine and had the money to buy it. But no, regulations are regulations and we had to provide at least two competitive quotations from another seller. The time for us to use the funds was coming to an end so we tried to find other vendors, but because the machine was very special (if I recall, it was a cryostat microtome), there was only one company (at that time) making that device. We found a couple of distributors that were selling basically the same machine, but at least in this manner we could gather two quotations. However, by the time we provided our purchasing department the various quotations and new forms duly signed, the time had expired and we could not use our money, hence we were unable to buy the machine. Here lies bureaucracy greatest triumph: to render the execution of even the simplest thing virtually impossible. These are just two examples of the impact of heavy duty administration in the suspension of research; I have many more that could fill out half of this book, but will stop here.

As a young student entering the university, one normally looks upon those very important and famous scientists with their white lab-coats with awe, thinking that they spend their days and possibly nights in the laboratory, painstakingly performing experiments and analysing data, asking questions of nature and receiving infrequent unambiguous answers, so they have to think and think again about another experiment, another inquiry that could unveil the phenomenon in question. Oh well, turns out those scientists are asking questions of another nature. It took me some time to realise this; it was perhaps not until I entered graduate school when I became enlightened on some of these matters… But I must admit I am not too bright!

Inspired by these and other events I lived through in the institutions where I worked, I sketched the following cartoons on the topic (Cartoon I). The last one where our friend Einstein is trying to obtain a PI position depicts a typical bureaucratic loop in science. Some of these loops are discussed in what follows.

CARTOON I

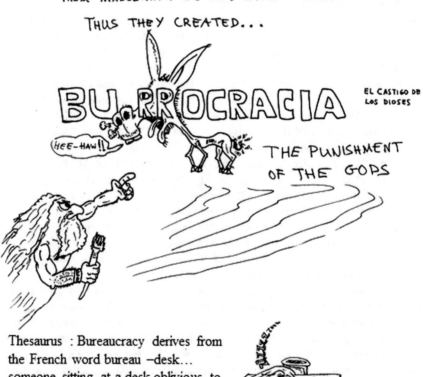

THE DEEDS OF HUMANS DID NOT GO UNNOTICED BY THE GODS IN THE HEAVENS, THUS THEY DECIDED TO PUNISH HUMANITY BY SENDING A PUNISHMENT THAT HAD TO BE NOT A QUICK AND FAST ONE BUT SLOW, TO LENGTHEN THE SUFFERING; NOT TRANSIENT BUT LONG-LASTING TO PROMOTE THEIR EMOTIONAL DISTRESS; NOT INTERMITTENT BUT EVER-PRESENT, TO DRIVE THEM MADDER... IF THIS COULD INDEED BE POSSIBLE.

THUS THEY CREATED...

BURROCRACIA   EL CASTIGO DE LOS DIOSES

HEE-HAW!!

THE PUNISHMENT OF THE GODS

Thesaurus : Bureaucracy derives from the French word bureau –desk... someone sitting at a desk oblivious to the needs and objectives of characters outside the office

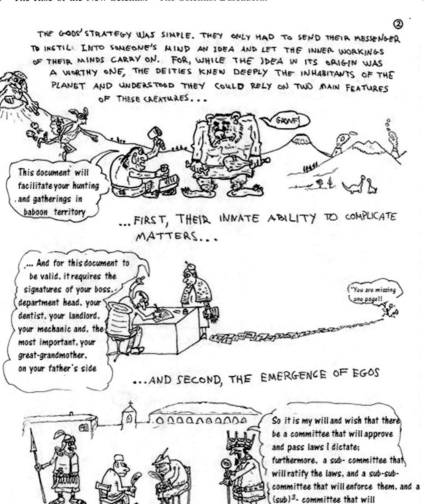

AND NATURALLY IT WAS JUST A MATTER OF TIME, ALLOWING THE
COMPLEX NON-LINEAR DYNAMICS OF THE SOCIETAL SYSTEM TO OPERATE,
THAT LOOPS DEVELOPED: FEEDBACK, FORWARD, FEED-FORWARD, RIGHT TO LEFT,
LEFT TO RIGHT, DOWN-UP, TOP-DOWN AND ALL POSSIBLE CLOSED LOOPS,
BUREAUCRACY FEEDS AND LIVES ON THEM...

...A STYLE OF THOUGHT THAT PROPAGATED AND INFECTED
OTHER INDIVIDUALS ORIGINALLY OUTSIDE THE BUREAUCRATIC
MACHINERY

1.1 Closed Loops

Bureaucracy generates loops, and many of them can be qualified as closed loops. It goes without saying that the invasion of administration in science has brought with it these "strange" loops. Depicted in the cartoon above (Cartoon I) is perhaps the most common, and perturbing, of the closed loops. Young fellows, do not be surprised when you apply for jobs to become independent scientists (PIs) if they ask you to demonstrate that you are able to write grants and thus obtain funding; problem is, as a trainee you cannot be a principal applicant in a grant (and normally not even a co-applicant): to apply for grants you need to be an independent scientist, need a job as PI. Hence a postdoc cannot write a grant (well, the fellow can write it, and in fact some PIs rely on their fellows to write considerable sections of grant applications, but will not be an applicant on that grant, as absurd as it may sound but that's the way it works) but to be offered the independent scientist position the institution wants to know you can write successful grants and, unfortunately, telling them that you contributed to grant writing during your postdoctoral rotation will not convince them: they want to see hard data, want to see you as applicant receiving the funds… Now you see the closed loop, one of many administrative nonsensical loops in this business.

Very related to this loop of demonstrating that one is able to write successful grant applications is another to demonstrate that you are creative and have your own ideas. To explain this, I will tell you what occurred to me during the time I was searching for my first job as a PI. The most common item that was brought up by almost all the institutions that interviewed me (if I recall, out of 8 interviews, the matter came up in 6 or 7) was whether I was truly an independent researcher, had my own ideas and develop, at least partially, my own projects. I say "partially" because as a postdoc —the step before applying to a PI job— you are working for the boss, the PI of the laboratory, who will put you to work in the project he/she decides. Depending on the laboratory, trainees may have more or less opportunities to develop their own ideas and projects; there are some very strict PIs who will not tolerate the trainee working on other research than the one dictated by him/herself. So you see, we have another potential loop developing: you are working for someone who tells you what to do and yet when you apply for independent positions they will require to know whether you have had your own ideas. To complicate matters, it is the PI of the laboratory who normally appears as the so-called senior, or last, author in the papers that trainees —the laboratory—publish. For those not in the field, let me clarify now that, at least in the biological sciences but I think in most of science and engineering, the authorship in a paper is determined thus: the first author is the one who has done most of the work, the last, or senior, author is the PI who has the money and —this being the crucial aspect for our current point— who is supposed to have had the main idea and designed the project, and in the middle go those who have worked in the project but to a less extent than the first author. Therefore postdocs are not normally last authors in papers and thus it is assumed they did not contribute much to the intellectual development of the project. How then was I to demonstrate to my interviewers that I had had my own ideas and developed my own projects during my postdoc time?

Without waiting for the next section about possible solutions, I will expound now a solution to these closed-loop conundrums. It all starts with the PI. I will explain what I did as PI in this regard. I allowed my trainees to work on their own ideas (besides, of course, working on the projects they were paid for) and in some papers where I did not contribute much, or nothing at all, I allowed them to be last authors; there were a few papers in which I did not even want my name as an author. But a tricky balance has to be reached, because first-author publications for trainees is a fundamental feature that will attract the attention of potential employers, hence it would be detrimental for a postdoc to be always senior author in publications, regardless of how creative and hard working the postdoc is. It is nonetheless advisable that PIs allow creative fellows to be last authors in publications where they have had a major intellectual contribution. These closed-loop troubles will continue to be diminished if, in addition to the aforesaid measures, the PI advisor writes a recommendation where it is clearly and without any trace of doubt stated the creative and independent nature of the postdoc. This is an aspect I always try to remark in my recommendations for my trainees, well, at least those who are sort of independent thinkers already. As an advisor, one must adapt to the trainees who come to the laboratory. There are those who want to be told what to do every day of the week. Fine, I have no problem with that. There are others who are more independent and have their own ideas for questions to be investigated, and they have to be allowed to develop these characteristics, they must learn how to ask proper questions and how to search for answers, that is, perform re-search. But now I am digressing, this book is not about being an advisor; however I would recommend advisors to think (hard) about the future of their fellows and the steps that have to be taken at some particular points in time to facilitate the opening of the several closed loops they will find when applying for jobs in academia. These closed loops here mentioned may not appear if the fellow applies for jobs in industry or in other spheres like editorial jobs, because in these positions creativity, independent nature or capacity to write grants may not be that fundamental as it is in research in academia.

One more inescapable loop occurs during grant applications. To have a successful grant that will be funded by the funding agency one needs to support the proposal with preliminary observations. But to obtain these pilot results one needs to perform research... Research that has not been funded yet, that is why one is requesting funds now. How to solve the loop? The typical action that we scientists do is to divert funds awarded to carry out another project to work on the new one and in this manner try to obtain the preliminary evidence that will demonstrate, without the shadow of a doubt, that the project will yield nice results. Is this redirection of moneys unfair? I do not know what to say, other than it seems the only possible manner to obtain a grant for a new project. And on the topic of being fair, let us acknowledge that there are institutions that offer money for pilot projects that in theory do not need preliminary data. But only in theory, at least such has been my experience with some of these "pilot project calls". The National Institutes of Health in the US has the programme called R21 —Exploratory/Developmental Research Grant Award— that in theory needs no pilot observations. Several years ago we submitted a R21 proposal and the requirement for substantial pilot data was made clear to me in the reviews, mind you, without mentioning the key word "preliminary data". The referees were concerned with the feasibility of the project, whether or not it would work, which is another manner of telling me they would have liked to see "strong" pilot observations that, again, will demonstrate without the shadow of a doubt that the project will yield nice results. I also met once someone who had served in one of the NIH panels, who told me that the instruction "No preliminary data is generally required" (as one can read in the NIH website describing types of grants) is… how to say… fictitious. When you have been many years playing the games in science —some of which will be revealed in sections below— you can accept it all without distress.

Those few loops here mentioned are what I would describe as the most perturbing loops in academic research. There are many more that are derived from the administrative culture, but to describe them would be too long and our space is limited. I will just finish by mentioning one of the last loops that I confronted, which was derived from the nature of my post as scientist in the two institutions where I worked, a hospital and a university (it is normal to have an affiliation to a university for those working in other institutions like hospitals or corporations). I inquired about the possibility of taking a sabbatical year, and therefore asked my Programme in the hospital, that directed me to the hospital Division where my position was held, but still being uncertain about my query, the Division asked me to inquire in my Department at the university, who, without answering yes or no, sent me to my Programme in the hospital to find the right response. The loop had been closed. I did not take the sabbatical.

Possible Solutions

Is there any way to escape the bureaucratic state? Undoubtedly, like the author of these lines, there are many who do not want to become trapped in administration, that want to remain scientists and not scientist-bureaucrats. In this section only some general recommendations will be offered, and will leave the many specific aspects related to this issue to following sections.

One first has to realise that these are the rules of the game and as such one must play it. Nevertheless, there are opportunities to minimise the bureaucratic burden. The first advice that comes to mind is to avoid importance; do not become too important, do not climb too high in the hierarchy. The more important one becomes, the more probable the researcher will be invited to join panels, committees, editorial boards etc. If your aim is to become a very high-ranking scientist —the head of a department, division or institution— you can be guaranteed beyond any reasonable doubt that you will step into your laboratory no more than two or three times per month, and only for a few minutes. On the other hand, if you are content with being a P.I. with a laboratory, if you are happy being not tremendously renowned, then you will most likely enjoy more time to devote yourself to that most satisfying of scientific demands: experimentation, the pleasure of directly asking nature a question about one particular phenomenon and finding out the response, or, as most of times go, trying to find out the solution, for research takes a long time and many experiments fail for one or another reason; but yet, the answer is waiting to be found there, and there is nothing more thrilling for some, like yours truly, that discerning that answer in situ, looking under the microscope, scrutinising a protein gel, or analysing the results of a computation.

But if —some readers will now be thinking— becoming an independent scientist, a PI, is the fact that determines when the problem starts, that is, when the scientist becomes an administrator, then the immediate remedy is not to become a PI. One can remain always a postdoctoral fellow or research associate and thus will be able to have much more time for experimentation and reflection. There are two drawbacks with this solution. First, postdoctoral fellowships are limited in time, after a while as a postdoc the incumbent has to change to either research associate (basically a higher type of postdoc), technician or, the normal thing to do, become an independent scientist — a PI. Second, not becoming a PI means one is always under the whim of someone else, the boss —the PI— who dictates the projects on which trainees and technicians work, hence one may never had the chance to investigate what one really wants to address, as it would depend on how open to suggestions the PI is and the freedom that trainees are bestowed in the laboratory.

In any event, not all scientists are like the one described two paragraphs above. Some enjoy administrative duties (although in my experience these are very few, yet they do exist), some want to climb very high in the academic ladder. And indeed it would be bad if no scientist was inclined towards administration, for nobody would want to become a head, a leader; but departments, universities and institutes need leaders and administrators. Some bureaucracy is necessary for academic organizations to function properly. Thus there is a necessity for the scientist-bureaucrat. The problem starts when this occupation is imposed on virtually everybody entering the ranks of professorship. Hans Wigzell, former president of the Karolinska Institute, believes that "there is a conflict between science and bureaucracy" even though "there is a need for good bureaucrats [...] the bad thing is when bureaucrats go into details and start micromanaging things following bureaucratic rules".

Young and aspiring researchers may find of interest to consult some blogs that are continuously appearing on the topic of how the current vast administrative infrastructure is changing what PIs should be doing —research— and what they are really doing, e.g. "Human Pipettes: Scientific training and education in biomedical research".

References

1. J. Johansson, Battling the bureaucracy hydra. Science **351**, 530 (2016). https://doi.org/10.1126/science.351.6272.530
2. http://www.sciencemag.org/careers/2016/09/why-aspiring-academics-should-do-less-science

Chapter 2
Time Is Precious—Quo Vadis, Creativity?

The previous chapter has served as an introduction of the global scenario that exists today in science and academia in general, scenario that will be further examined into its particular facets in the following chapters. Scientific research that leads to clear results relies on obtaining large numbers of experimental observations and reflecting in depth about those observations and associated theories and hypotheses. This, however, is becoming unfeasible in current times due to the lack of time and funds. The former, lack of time, in my opinion is most troublesome, because there is research that needs not much money (e.g. theoretical investigations) but all research, cheap or expensive, needs time.

CARTOON II

We have created a world in which time is a commodity (Cartoon II). With regards to the scientific enterprise, the many administrative responsibilities that scientists have in their hands create the dilemma of how to allocate time to the diverse demands they are supposed to carry out —including research! But the time predicament is a common aspect in modern societies in general, and specifically

© Springer Nature Switzerland AG 2019   15
J. L. Perez Velazquez, *The Rise of the Scientist-Bureaucrat*,
https://doi.org/10.1007/978-3-030-12326-0_2

today's scientists have to face it in manners that remove many aspects of true scientific research. One of the casualties is proper thinking and reflection. The words of Bruce Alberts and colleagues illustrate perfectly the situation: "Today, time for reflection is a disappearing luxury for the scientific community" [1]. These authors are talking specifically about the time a researcher needs to think and reflect on the experiments to be done and on the interpretation of the observations; thinking is the foremost activity of a scientist, and yet today it is becoming a luxury. Many experiments, countless projects today are done without much reflection. It is not uncommon that one scientist, upon reading about experiments done in a certain model system —say for example electrophysiological recordings done in neurons— will try to repeat similar experiments in his/her favourite system, perhaps muscle cells. It is not only the scarcity of time to think but also the pressure to publish many papers and to obtain the so-called positive results which favours the annihilation of truly creative thought and forces scientists to use strategies that will accelerate their productivity in terms of the aforementioned results, papers and grants.

There is another common strategy to design a project that bypasses much cogitation. Pick up a few scientific journals, find out what the hot topic is, and join the club. So, if in biochemistry a protein kinase is creating passion among biochemists for one or another reason, then it is conceivable that if you start working on that protein you will obtain funds more easily and publish in better journals, because this is the hot topic of the time. During my career I have witnessed innumerable hot subject matters that appear and, after a while (sometimes a very short while), vanish. But regardless of the evanescent nature of these molecules, planets, or chemical compounds, one could profit from them while they are of supreme interest to scientists, corporations or governments. Having said all this, it is true that one still has to have enough acumen to devise a project on that hot topic, but the main difficulty, so to speak, of finding out what problem to tackle, has been avoided as it was handed by the current trends in research.

We always have to keep in mind that these behaviours —and many others that will be discussed in other sections— are due to the nature of the world of science in these days, and, in addition to the time issue, they can be traced back to one common source of problems: the extreme competitive environment that has derived from the scarcity of funds for research. You see, in olden times there were fewer scientists —much less money too but enough for the few scholars to mind their business— and therefore the competition, that always existed, was not as intense as it is today. Some authors have used terms like hypercompetition, and have devoted many words to demonstrate and spread awareness about how damaging this situation is for science. There will be a chapter below devoted to the never before seen contemporary tremendous competition in science.

At the same time, it is fair to say that those who care less about funds and about where their papers are published may not succumb to these behaviours. It is, after all, a personal choice. There are some who are in science to gain fame and fortune. Others just to obtain a job with a salary. Yet there are others who just want to satisfy their curiosity about certain natural phenomena, regardless the importance society gives to those phenomena. It is, in the final analysis, a personal choice. Of course if one is lucky may find that one's main interest lies within the boundaries of extreme societal importance —like some research that has very practical applications to aspects of healthcare— and then it is possible to become renown and still follow one's passion.

Nevertheless, there are important consequences of the lack of time scholars face these days. Apart from the loss of creative thought, a fundamental consequence that is creating tribulation is the urgency in performing experiments to obtain results, which inevitably ends up in the publication of flawed results that underlies the so-called decline effect.

2.1   Irreproducibility Issue: The Decline Effect

The "decline effect" merits some comments, as many lay people tend to believe — almost to the point of faith— in scientific results that are announced in the media. This decline phenomenon refers to a decreasing support over time of "positive" scientific results. Just to be clear, a positive result is one which supports the proof of a theory or idea; negative results are those that either disprove it or do not add anything clear. The unfortunate situation is that only "positive" results are published (but see the section below, where you will find a few journals devoted to the publication of negative findings), while the others tend to be ignored; will come back to this later.

The decline effect implications are troublesome particularly as it occurs in treatments used in health care. Before being released to the public, drugs and other medical treatments undergo clinical trials. If the therapy demonstrates good, "positive" results in a patient population during the trial, then it will be eventually commercialised and, if it is a drug, you will find it in the pharmacy. Hence, companies that design drugs and medical therapies are very keen to see positive results in their clinical trials. These trials involve many subjects, and require, normally, substantial funding. They are done in strictly rigorous manner, at least according to the current gold standards in clinical trials, which there is no need to describe now. Suffice to say that these experiments are very carefully done and evaluated. It has to be so, as the health of the population depends on the therapy and nobody wants to release a harming product. The article "The truth wears off" by J. Lehrer (*The New Yorker*, December 30th, 2010), that tries to address the nature of the decline phenomenon, starts by describing the bewilderment among the participants in a meeting where it was announced that the therapeutic efficacy of some antipsychotics was, apparently, fading, even though previous several clinical trials involving thousands of subjects showed excellent effectiveness at decreasing psychiatric symptoms. This is, in essence, the decline effect: initial excellent positive results are not replicated in later experiments. It applies to large clinical trials and small laboratory projects.

To be honest, I must admit there is something strange regarding the experiments that lead to the initial observations of a phenomenon. It has been my own experience, and that of other colleagues, the initial prevalence of positive results, those that support the theory or hypothesis. As odd or even esoteric as it may sound, many times my colleagues and I in the laboratory have been confronted by the fact that the initial set of experiments yield such good results that forced us to continue doing more, only to find out, a few days later (or, depending on the nature of the experiment, sometimes even a few hours later), that the first results were not reproducible. In short, it was a fluke. The weird thing is that it happened, to me at least, a good number of times; too bad, because we wasted time investigating things that led nowhere. Under the assumption that experimental observations are distributed along a Gaussian —also called normal— distribution, then the aforementioned situation of finding very interesting results means that the initial experiments provided the observations that were located in one extreme of the Gaussian. Those are the results that are "positive" in that the significance of finding such outcomes is high; to assess significance many statistical tests are used, starting with the simplest one, the Student's t-test, and normally a significance value "p" is reported in papers. The observations that lie in the extremes, or tails, of a Gaussian are those of high significance, or equivalently low p values. If one starts making observations of a phenomenon, those more numerous, and thus of low significance, are located towards the centre of the Gaussian curve, therefore these results should be those initially found when starting the collection of observations, just because they are many. If you throw five dice you will find that extremely very few times the five will show the same number; the most numerous outcome, therefore the most probable, is that the five faces are different. So imagine you start your dice experiment, cast the five dice 6 times in a row and you observe that in five of those six times the dice have shown same number. That is very interesting, because it is so improbable, and prompts you to continue gathering observations, thinking something worth investigating is going on. Then, as you proceed casting the dice, you observe, not without certain disappointment, that the next 10 times the results are the expected: a combination of faces of the dice. This is nothing more than the regression to the mean, as more observations are collected early statistical coincidences are cancelled out. This is the decline effect in the very tiny scale of the dice experiment. But, as aforementioned, this initial finding of interesting results occurs quite often, as though the first attempts were probing the tails of the Gaussian distribution (those of low probability), and not, as it should be, the middle of the distribution (because the probability is larger). This digression on statistics has the purpose of illustrating that the decline effect may not be as surprising as one may think. Yet, in case of small scale laboratory experiments we are talking about a few samples, but in large clinical trials there are thousands of data (each patient counts as one observation). Hence, in the former case it may not be too extraordinary that the initially collected values are located in the extreme of the distribution, but in the latter cases, with so many observations, it looks bizarre. Thus sometimes the initially positive results are not reproducible under more data collection.

2.2 On the Nature of Irreproducibility

But, let us start by asking the obvious: should scientific results be reproducible? From very large clinical trials to small-scale experiments, the failure to replicate original results seems commonplace, occurring in any discipline be it parapsychology or medicine. Indeed, it is the irreproducibility what appears to be reproducible! Yet, reproducibility of results is the fundamental cornerstone, the very foundation of scientific research: reproducible experiments in a controlled experimental situation. In my opinion, this statement is, in many cases, a fallacy. This is not said as a challenge to science, rather as an explanation of the mysterious decline effect. Let us, I beg you, consider the nature of different experiments that are commonly done. Some experiments are carried out at equilibrium conditions (these considerations may require a basic knowledge of thermodynamics, so I excuse myself aforehand), say, the typical binding assay in biochemistry that is allowed to reach thermodynamic equilibrium: the test tube that contains the reactants is place on ice for some minutes so that chemical equilibrium is reached. Because of the nature of chemical equilibrium (there is no need to delve into chemistry now), the data will look very much the same time after time the binding assay is done. This is the beauty of systems at thermodynamic equilibrium, they provide results that are highly reproducible. Other experiments that require close to equilibrium conditions may be, in fact, quite reproducible. But, at least in life sciences, most experimental situations occur in far-from-equilibrium situations. As it is well known to physicists and other scientists, far from equilibrium is the source of the richness in our universe, it is the basis of the recipe for interesting things to occur, to wit, turbulence, chaotic dynamics, fluctuations, and thus the variability that originate complex phenomena (for those interested, a not too specialised review is Jaeger and Liu's article "Far-from-equilibrium physics: an overview" [2]. Because of this lack of equilibrium with its consequences, the notion of controllability in the research conducted is a fallacy in these conditions and thus irreproducibility follows naturally.

An example will illustrate the fallacy of our presumed "controlled experimental conditions". A typical experiment done in neurophysiology is to record electrical activity in brain slices. I have chosen this experiment, in vitro electrophysiological recordings, because it is close to what should be equilibrium conditions, at least for the brain slice sitting in the recording chamber. The slice is normally superfused with some medium that mimics the cerebrospinal fluids in the body. So one puts the tiny slice in a recording chamber, starts superfusing with the medium and waits a few minutes until the temperature and chemicals in the chamber may have reached equilibrium. Then one starts the electrophysiological recordings, after the experimental set-up is nicely controlled... But, is it? First let us look at the recording set-up. There could be differences in temperature from one day to another that will result in the medium to start flowing at different temperature. Indeed, it occurred to us that some data gathered in winter looked different from that collected in summer, and one of my astute colleagues realised that because the room where the electrophysiological set-up was located did not have air conditioning, the temperature in winter was much lower than during summer time, hence the medium started flowing at different temperature depending on the season when the experiment was carried out. It is true that the recording chamber is set at a specific temperature with the help of a thermostat, but if you run liquid of different temperatures there will not be enough time for it to equilibrate because the fluid runs very fast via tiny tubes that discharge the medium into the small recording chamber. Our problem was fixed by warming up the medium always at the same temperature before the experiment... But yet it will cool down with time during the long experiment that may last hours. As another source of daily variability in the experimental conditions, the tubes may support the growth of mold or other microorganisms, and these release substances that will end up in the recording chamber and may alter the physiology of the sample. This is why, as all neurophysiologist trainees are taught, cleaning the set-up after use is as important as the experiment itself. Then there are days when, mysteriously, electrical noise appears and affects the recordings; sometimes the source is found, other times is not. Anyhow, as one can see, experimental conditions may not be as controllable as one may think and require time and effort to achieve that desired "control". And then we have to consider the brain slice itself, a piece of living material and thus not at equilibrium (equilibrium for the living means death). We assume the chemicals reach equilibrium but this occurs only in the chamber, and not to the same degree in the tiny interstices of the living substance. In sum, all these events will influence the recording.

The previous example showed that even in conditions that are relatively simple and "clean", a perfect control of the experimental conditions may not be possible. The situation worsens considerably when one talks about, say, cognitive experiments, or in vivo electrophysiological recordings. There is neither space nor time to provide illustrations of some typical experiments in these fields, let us just say that one has to do the humanly possible to try to keep constant the conditions, and hope for the best. In neurocognitive experiments, with myriad of factors at play—starting with the mood of the individual who volunteers for the experiment— a well-controlled situation is nearly unfeasible. This may be the reason why there is great variability in the results reported in many cognitive projects. A good illustration is the paper by Cabeza and Nyberg "Imaging cognition II: An empirical review of

275 PET and fMRI studies" [3] that examines the variability in 275 neurocognitive experiments. To the experimental conditions we have to add the immense variability in the analytical methods used to study the recordings and data gathered. J. Carp, in a paper on a systematic review of 241 functional magnetic resonance imaging studies [4], showed that there were almost as many analytical pipelines as there were studies (241!). If various analytical methods are applied to high-dimensional data that are typical of cognitive science, the generation of false-positive findings is likely and therefore variability in results is a natural consequence.

The preceding, perhaps too extensive, narrative on aspects of the notion of reproducibility in research indicates that in many projects, particularly related to life sciences and others where the systems are far from equilibrium, variability of results is natural and to be expected. Therefore, variability should not be a declared enemy of scientists, rather, a natural consequence of nature's dynamics. Nevertheless, there is concern about reproducibility; as examples take a look at the editorial "Is there a reproducibility crisis? [5] and at the perspective article "A manifesto for repro-ducible science" [6]. For those who want more, Nature, one of the multidisciplinary journals, has a special collection of editorials and articles on the matter of repro-ducibility at https://www.nature.com/collections/prbfkwmwvz ('Challenges in irreproducible research').

2.3 Positive and Negative, What Is Published and What Is Not

I have not failed 10,000 times. I have not failed once. I have succeeded in proving that those 10,000 ways will not work

Thomas A. Edison

The only way to know what the correct decision is, is to know which is the wrong decision

Paulo Coelho, *O Diário de um Mago*

The roots of the problem —if we so decide to call this state of affairs— about the reproducibility issue and the associated decline effect include two main factors: the predisposition towards publication of "positive" results, as was briefly mentioned above, and the emphasis that is placed on statistical comparisons along with some misconceptions about statistical significance values. There is substantial concern too about the trend to only publish positive results, and the editorials and literature on the topic is increasing steadily. It may be misleading, perhaps a misnomer, to use the term "negative results", as these are as useful as the positive ones. Thomas A. Edison put it very clearly: "Negative results are just what I want. They're just as valuable to me as positive results. I can never find the thing that does the job best until I find the ones that don't."

Solutions have been proposed. For one thing, there are even journals devoted to publication of negative results: 'Journal of Negative Results in BioMedicine', 'Journal of Negative and No Positive Results', 'Journal of Negative Results',

'Positively Negative' which is a PLOS ONE collection focusing on negative and inconclusive observations, and F1000Research. All articles benefit from transparent refereeing and the inclusion of all source data. Other advanced solutions include the creation of open access repositories, see for example Schooler's article "Unpublished results hide the decline effect" [7].

But the need to acknowledge and publish negative results is easier said than done. First, this entails a paradigm shift in that "negative" results should carry equal weight as "positive" ones. But it may not suffice. Second, a change of habits in data presentation and interpretation could be fundamental too. As always occurs in life, one event is related to another, nothing stands alone. Thus, we have seen that the reproducibility subject involves several aspects of the scientific enterprise including ignorance of negative results and the assumed controllability of experimental set-ups, and to these one must add at least a couple more that are very connected too: the presentation of data and the lucrative aspect of the positive, very interesting, results.

2.4   Life of p: The Misrepresentation of Reality

My deeper motivation is a feeling that numerical exactitude is alien to the diversity of organic evolution, and pretence of exactitude often obscures the qualitative essentials that I find more meaningful

Arthur T. Winfree, *The Geometry of Biological Time*

If one looks at papers written today, the immense majority of them —in the realm of experimental sciences— are full of p values that denote statistical significance. Data are usually presented as neat averages with tight standard deviations and p values. Scientists are basically chasing statistical significance and ignore the results that do not have low enough p (the lower p is, the more significant the observation is), the very valuable —in my opinion and that of T. A. Edison as shown in his words in the previous paragraphs— negative results.

There is nothing wrong presenting grand averages, means and standard deviations. But the presentation of average values conceal the variability, and this is why it is the preferred means to present data; we all like our data to be strong, clear, convincing, and showing variability does not help. Averages do not show the whole picture, which is represented by the complete probability distribution, where one can see the tails of the distribution, the outliers, that are as well possible results one may obtain doing the experiment. Perhaps if you are a scientist you have sometimes tried to reproduce some published experiments that you need to start a project. And perhaps sometimes you could not reproduce the data as shown in the papers, because occasionally your initial observations will be precisely the outliers, whose values will be far from the average presented in the articles. As you keep doing more experiments the observations will regress to that mean value published in the papers you chose as starting point of your project, but it could be discouraging that the first few data gathered are far from the mean, and, frequently, one stops doing those experiments: end of the project before it started!

Hence, in addition to presenting averages plus minus standard deviations and associated p values, it could also be very useful to show the full probability distribution function (pdf). The view of the entire pdf will help envisage what can be expected when performing those experiments: the outliers are clearly visible in the full distribution (and quite often, those outliers are of fundamental importance), thus revealing the expected inconsistency. A brief look at any pdf gives one a more comprehensive, global perspective of the phenomenon. Hence when you start collecting preliminary data you may not be that discouraged, because you know your observations are part of the tails of the pdf and thus possible, albeit improbable. At the same time, as an additional benefit of displaying the complete pdf, there will be a lessening of the absolute emphasis that is placed on statistical comparisons. Sir Francis Galton articulated his views on the topic in the book Natural Inheritance, written in 1889: "It is difficult to understand why statisticians limit their inquiries to averages and do not revel in more comprehensive views. Their souls seem as dull to the charm of variety as that of the native of one of our flat English counties, whose retrospect of Switzerland was that, if its mountains could be thrown into its lakes, two nuisances would be got rid of at once".

This is not a treatise on statistics, thus we will not go into what p values really mean and why many times they are misused and misunderstood. All graduate and many undergraduate students are indoctrinated in the importance of statistical significance, but I gather based on my own experience and what I read in some articles and editorials, that very few really comprehend what the term represents. It would be equally important that students are taught the origins of statistical significance, especially the story of (Sir) Ronald A. Fisher who very arbitrarily declared the idea that if a number called p (after probability) is less than 0.05, then the result is statistically significant. Today, more than a century later, his whim, this arbitrariness, rules the world of science and medicine.

The importance placed on statistics and the dictatorship it exerts could be misplaced and misleading, and I find reasons to doubt that data presentation and interpretation should be dictated by an area that we think is as solid as was, e.g., Euclidean geometry at one time. The current faith on power estimations and other statistical schemes should be seriously revised. Numerous works have addressed the subject of the misleading notion of statistical significance (e.g. *"The Cult of Statistical Significance: How the Standard Error Costs Us Jobs, Justice, and Lives"* [8]). The situation is such that even the American Statistical Association released a policy statement in 2016 noting that one of the major contributors to the irreproducibility "crisis" is the misuse and misrepresentation in research of what p denotes, and proposes a set of principles to guide a reasonable interpretation of statistical significance values [9]; the main message of this editorial is that scientists should avoid drawing definite conclusions based on values of p. There are those who think that even the terms 'significant' or 'non-significant' should not be used in reporting data. The misuse of p is clearly troublesome in medicine, as clinical trials are (mis) using these statistical notions to approve therapies ([10], and in Ref. [11] you will learn "Why most published research findings are false").

The obsession with excessive statistics and quantification is perhaps driving us to forget that science is a process of asking questions —normally about natural phenomena— and finding answers and not a rigid sequence of numbers that quantify everything that is quantifiable. Perchance we are losing the capacity to ask questions. Ronald D. Vale may have a point when he declares, in his article "The value of asking questions" [12], that "Science begins by asking questions and then seeking answers. Young children understand this intuitively as they explore and try to make sense of their surroundings. However, science education focuses upon the end game of "facts" rather than the exploratory root of the scientific process. Encouraging questioning helps to bring the true spirit of science into our educational system". I agree with him. In my opinion, teaching children and youths the art of asking proper questions and moderating the fixation with exactitude (as A. Winfree declared in his words reprinted at the start of this section), explaining to them that all we see in nature are tendencies rather than absolute facts or certainties, will improve the course of science.

Related to this subject matter, we find today that much of the emphasis is not placed on question-driven projects, but rather on hypothesis-driven proposals. Once again, we have to be fair and acknowledge that both of these approaches to develop a research programme are acceptable, but the problem today, as I and many others see it, is that funding agencies insist on hypothesis-driven proposals. You have a much better chance to have your grant funded if you start with a clear hypothesis the referees can comprehend. I know very well that some scientists will not consider a proposal unless an unambiguous hypothesis is delineated. It would be more reasonable and fair to dispense with this mind-set so that both strategies are equally considered, and it becomes even more advisable if we consider that many great scientific and technological advances were not driven by any hypothesis, to wit, the works of Newton, Galileo, Darwin, Crick, Einstein, Boltzmann and many others were not in desperate need for hypotheses. Rather, it was a mixture of intuition, appropriate questions and applied logic what defined most of the great scientists' endeavours. The inventor and cognitive scientist Marvin Minsky said in his "Memoir on inventing the confocal scanning microscope" [13]: "I've never had much conscious sense of making careful, long range plans, but have simply worked from day to day without keeping notes or schedules, or writing down the things I did". I find it refreshing to read his words… but his working habits may not work for many of us, very limited in our intellect, who need to keep careful track of our endeavours.

To conclude, then, do the decline effect and the related variability in experimental results really represent a problem? One of humanity's greatest talent is the creation of problems where there is none. Yet, we must admit that it is a problem for the experimentalist that has difficulty replicating results due to the current nature in data presentation. On the other hand, it may not be a genuine problem, but an expression of our present-day ignorance and assumptions, essentially the aforesaid supposition of predictable equilibrium states in our experimental conditions. When I read papers, I almost always believe the results, for after all, the observations and related measurements are real, but I also recognise that these may amount to close to nothing— or may be misrepresented or misunderstood, that's another matter —perhaps just indications of a tendency in some particular phenomenon. Because, and this is a fundamental notion that, it is my opinion, should be engraved in all students heads, it is only *tendencies* and not essential truths that we observe when performing our laboratory experiments. Thus, this irreproducibility crisis is resolved and our minds put to rest if we accept the fact that we discover tendencies and not certainties (let me insist again). And the reason that scientists are limited to basically discover only tendencies is found in the fabric of nature: the dynamics in most of natural phenomena are governed by metastability, that is, a form of instability from where variability emerges: stochasticity, randomness, noise, whatever you call it, is thus a perfectly natural state. This concept is perfectly expounded in Kelso and Engstrøm's book *The Complementary Nature* [14]. Therefore, accepting the intrinsic dynamic instability of natural phenomena in general, one concludes that there is nothing wrong with experimental results. Stochasticity is present at all levels of description, from sets comprised of data points to sets constituted by reported results. Stochasticity is accepted in sets of data points that constitute the observations done in experiments; why should not it be accepted in sets of individual experiments (imagine that each data point in such a set is a published paper rather than a specific measurement or observation)?

Pondering on these matters, one could think that the situation related to data presentation represents a version of the famous prisoner's dilemma (Cartoon III).

CARTOON III

Have you ever tried to reproduce a published experiment and find that you can hardly obtain the previously published results? Then learn about the prisoner's dilemma, for we scientists are playing it. The tendency of ignoring the "outliers" in the presentation of results is as common as that of eating breakfast upon waking. This tendency did not appear by chance, but forced by the scientific review process, and, not surprisingly, underlying this pre-disposition we find the competition for funds, positions, and glory. Thus we are playing a variant of the celebrated prisoner's dilemma in game theory, and we are afraid this is leading to trouble. The prisoner's dilemma shows why two completely rational individuals might not cooperate, even if it appears that it is in their best interests to do so. For those who do not know it, this is the basics (taken from Wikipedia).

"Two members of a criminal gang are arrested and imprisoned. Each prisoner is in solitary confinement with no means of communicating with the other. The prosecutors lack sufficient evidence to convict the pair on the principal charge. They hope to get both sentenced to a year in prison on a lesser charge. Simultaneously, the prosecutors offer each prisoner a bargain. Each prisoner is given the opportunity either to: betray the other by testifying that the other committed the crime, or to cooperate with the other by remaining silent. The offer is:

- If A and B each betray the other, each of them serves 2 years in prison
- If A betrays B but B remains silent, A will be set free and B will serve 3 years in prison (and vice versa)
- If A and B both remain silent, both of them will only serve 1 year in prison (on the lesser charge)

Because betraying a partner offers a greater reward than cooperating with them, all purely rational self-interested prisoners will betray the other, meaning the only possible outcome for two purely rational prisoners is for them to betray each other. The interesting part of this result is that pursuing individual reward logically leads both of the prisoners to betray when they would get a better reward if they both cooperate and kept silent."

Applied to scientists in the particular issue of data presentation, it can be conceived that cooperation would imply that everybody presents all data clearly so that the variability or negative and neutral results are evident, while defection means the concealment of data points that are outliers, the sup-pression of variability and the presentation of only positive results. Indeed, more than once have I experienced the result of presenting, in papers sub-mitted for review, the variability inherent in experiments, only to be asked by reviewers either not to mention it or to remove those inconvenient data points that complicate the interpretation of results. Since revealing inconsistency normally results in problematic acceptance of papers, scientists thus defect, and rarely do we see comments on the ambiguity and irregularity of experimental manipulations, for we all know that by ignoring rare events that distort nice and consistent averages, our papers will have better chance of publication that will allow us to successfully apply for funds and obtain full-time positions. If we were to cooperate by acknowledging in all honesty the variability in results, it would make things easier when trying to reproduce specific experiments because the experimentalist would have a clear idea of the intrinsic irregularities in the expected results. However, much like in the original prisoner's problem, the possible defectors who would choose to ignore variability would obtain almost perfect results, to the reviewers' delight, and thus may gain advantage by publishing more articles than the co-operators. Hence cooperation seems to be a rather unstable state and in the end defection would spread, notwithstanding the aforementioned problematic consequences of reproducibility. Cooperation in this particular matter would have important consequences not only in the practical sense of performing experiments knowing the inconsistency we may expect and thus making our experimental lives easier, but also for the understanding of natural phenomena, for quite often those outliers are of fundamental importance, rather than the neat averages with tight standard deviations. Just ask seismologists, could they do their job if they were to throw away the big earthquakes that represent absolute outliers when compared with the myriad of smaller and unnoticeable (unless you have the right equipment) seismic activities? What do you advise then our cartoon character above to do?

References

1. B. Alberts et al., Rescuing US biomedical research from its systemic flaws. PNAS **111**, 5773–5777 (2014). https://doi.org/10.1073/pnas.1404402111
2. H.M. Jaeger, A.J. Liu (2010) Far-from-equilibrium physics: an overview. arXiv:1009.4874
3. R. Cabeza, L. Nyberg, Imaging cognition II: An empirical review of 275 PET and fMRI studies. J Cogn Neurosci **12**(1), 1–47 (2000)
4. J. Carp, The secret lives of experiments: methods reporting in the fMRI literature. Neuroimage **63**, 289–300 (2012). https://doi.org/10.1016/j.neuroimage.2012.07.004
5. M. Baker, 1,500 scientists lift the lid on reproducibility. Nature **533**, 452–454 (2016)
6. Munafò et al., A manifesto for reproducible science. Nature Human Behaviour **1**, 0021 (2017). https://doi.org/10.1038/s41562-016-0021
7. J. Schooler (2011) Unpublished results hide the decline effect. *Nature* 470, 437. https://doi.org/10.1038/470437a
8. S.T. Ziliak, D.N. McCloskey (2008) *The Cult of Statistical Significance: How the Standard Error Costs Us Jobs, Justice, and Lives.* The University of Michigan Press

9. R.L. Wasserstein, N.A. Lazar, The ASA's statement on p-values: context, process, and purpose. The American Statistician **70**(2), 129–133 (2016). https://doi.org/10.1080/00031305.2016.1154108
10. J.A.C. Sterne, G.D. Smith, Sifting the evidence: what's wrong with significance tests? BMJ **322**, 226–231 (2001)
11. J.P.A. Ioannidis, Why most published research findings are false. PLoS Medicine **2**(8), e124 (2005)
12. R.D. Vale, The value of asking questions. Mol Biol Cell **24**(6), 680–682 (2013). https://doi.org/10.1091/mbc.E12-09-0660
13. M. Minsky, Memoir on inventing the confocal scanning microscope. Scanning **10**, 128–138 (1988). https://doi.org/10.1002/sca.4950100403
14. J.A.S. Kelso, D.A. Engstrøm (2006) *The Complementary Nature,* MIT Press

Chapter 3
Corporate Culture in Academia and The Current Standards of Research Appraisal

One sees evidence of scientists spending money instead of thoughts [...] converting university professors into administrators

Alvin M. Weinberg, 1961

Along with the advent of the scientist-bureaucrat has arrived the rise of academic capitalism; in fact, one aspect is a reflection of the other. It should not be surprising that sooner or later money becomes one main objective and product of scholastic development. We only have to remember that long, long time ago, lucrative greed entered even the most sacrosanct activities, like the Olympic games, where Greek athletes went from competing for purely sportive reasons to compete for the money paid by the Roman conquerors of Greece. Money talks, they say. It has always been talking, except perhaps in very ancient times when hunting-gathering kept humans very busy. Going back to science, this situation has resulted in the emergence of new standards by which scientists are judged. It is fair to evaluate scientists' activities from time to time, but in modern times the appraisal of research has reached such proportions that inflicts more limitations —as if there were few already— to creativity, to the basic activity of scientists: innovative thinking.

I have been lucky to practice research in a time when universities were not corporate organizations yet, but they are becoming today. To a large extent, I have been able to investigate matters that interested me, regardless of financial retribution to my organization or possible instant application to healthcare. In short, I followed my vocation. It is important to recognise that universities were created, in times of old, to teach and investigate, and were largely independent of policy makers, the scholars being rarely evaluated for their performance as it was assumed they knew what they were doing. But in the past few years the researcher's vocational autonomy is being replaced with management and external control by the institutions.

© Springer Nature Switzerland AG 2019
J. L. Perez Velazquez, *The Rise of the Scientist-Bureaucrat*,
https://doi.org/10.1007/978-3-030-12326-0_3

3.1   Avoiding Risk, Funding Triviality—The Infinitesimal Approach to Research

The unexpected is the whole point of scientific research

Brian Flowers

New ideas pass through three periods: it can't be done; it probably can be done, but it's not worth doing; I knew it was a good idea all along

Arthur C. Clarke

To obtain enough funds to maintain a laboratory, a scientist today needs to perform utilitarian research. Try to submit a grant proposal with a project that does not deliver, apparently, an almost immediate profitable avenue, and most likely your project will not be funded. To be fair, it is true that there are few organizations that fund pure research, that which is driven by curiosity and tries to comprehend and characterise natural phenomena regardless of the lucrative aspect. For instance, in Canada there exists the Natural Science and Engineering Research Council (NSERC) that will fund truly academic research, in addition to more applied projects. But look at the grants given by the NSERC as compared to, say, the other Canadian institution that funds biological research —that which is evident that will have a direct impact in healthcare— the Canadian Institutes of Health Research (CIHR): a normal operating grant awarded by the former institution brings around $35,000/year (Canadian dollars), while the CIHR average grant offers about $170,000/year (in 2017), almost 5 times more. This Canadian example can be easily translated to almost any other country in the world. All types of research should be funded, but looks like there is not much of an equilibrium, a balance between funds awarded to "pure research" and those given to applied science. The unquestionable fact today is that granting agencies favour utilitarian research, there is an excessive emphasis on short-term research that can be translated into commercializable products. Not surprisingly, 'translational research' is a major keyword today in science. And all this despite the fact that, in the end, theoretical research may end up having extraordinary applicability; governments and administrators should learn that the potential practicality of all types of research should never be dismissed.

The fulfilment of the vision of science may be the understanding of natural phenomena, but in current times one has to be sure the phenomenon in which you are interested has a practical application in industry, healthcare, business etc. Otherwise, you will obtain little funds and, unless your research is cheap, you will not be able to carry out experiments. As always, an example is worth many words. Let me tell you our own experience working in the field of traumatic brain injury. Initially we started performing biochemical and other experiments at the molecular/cellular level that investigated possible biochemical targets to ameliorate damage caused by brain trauma and ischemic injuries (like those occurring after strokes), and we were able to obtain substantial grants that funded experiments and several salaries. This research had, obviously, potential pharmacological application, thus attracted attention of funding bodies. As it happened, I started to become more interested in studying brain dynamics after trauma, this research done not at the cellular level as the former, but at the so-called systems level. It is a more, should we say, theoretical study aimed at understanding the activity of brain cell networks after brain injuries. Therefore we sought to obtain funds to carry out our new line of research, mind you, in the same field: traumatic brain injury. The difference was that this research at the systems level did not have a very immediate, practical application, that it was more pure research than the one we developed studying molecules involved in traumatic injury. Year after year we submitted grant applications to agencies in Canada and beyond; I recall at least 6 funding organizations, and at least same number of grants. We were not able to obtain a single cent. To be fair, one of our students was indeed able to obtain a scholarship to carry out studies in this area of brain dynamics after trauma, but as for operating grants, nothing at all. Thus one can see that working in the same field but changing levels of description, from the molecular/cellular to the systems level, brought about the impoverishment of the laboratory. The truth is that the research we conducted on brain dynamics could be extremely useful in the clinic, but at that time it was the beginning of this type of research in the brain trauma field and thus it was more adventurous and sounded too theoretical. Similarly, in the theme of brain dynamics in epilepsy, we were able to obtain only one grant funded out of more than 8 attempts in the period 2007–2015.

Funding institutions do not like to take risks, do not appreciate uncertainty, and thus will almost never fund truly creative, innovative research, because, as a consequence of being creative, this research is risky: we do not know what the results will be. Real research is therefore inherently uncertain, it consists in investigating the unknown; if we knew the results that will be observed it would not be called research —someone said once— and therefore the conclusion seems to be that granting agencies do not fund research. What do they fund? What is mostly funded nowadays is what Thomas Kuhn mentioned in his book *The Structure of Scientific Revolutions* (1962), his opinion being that scientific progress is advanced by accumulation of accepted facts and theories; so what is funded in our time are projects that seek to uncover details about phenomena that are more or less well characterized, mainly projects that aim at confirmation and clarification of already known findings. And the more trivial, easy to produce "positive" results the details are, the more likely the project will be funded. Kuhn's assessment of "development-by-accumulation" of accepted facts has been the standard since the middle of the past century, and possibly before that. It is no coincidence that Gerhard Fröhlich is credited with the words: "Most scientific publications are utterly redundant, mere qualitative 'productivity'".

Not that there is something wrong with the investigation of details about specific phenomena, for science evolves taking little steps; sometimes infinitesimal steps. The cause of trouble is when there is no balance between the funding of the infinitesimal approach and the more holistic, adventurous projects. This lack of balance pervades many aspects of research, as it is becoming apparent in what I am bringing up in this book: imbalance in funding big groups versus the individual, disparity in funds for infrastructure and for operating research, inequality in the scientists' time dedicated for administrative work versus real research in the laboratory, disproportion, inequity in the appraisal of research favouring the administrative and lucrative outcomes over work done in the laboratory… Indeed, I think the general situation in academia and science in particular would be almost fixed (if not all, at least many facets of the current situation) if an equilibrium was found in all these, and perhaps others, aspects.

In the case in point now, the following is the normal course of grant application submission and resubmission: one starts writing the proposal describing the original scheme of the project as it was envisaged in its conception, then some time after submission to the agency the referees' criticisms and scores are received resulting in the grant not funded as it stands, and one reads that the project has to be more focussed on particular aspects; therefore the proposal is re-written leaving more holistic features aside and upon resubmission one may find that the level of detail was still not enough, hence one more round of writing and submission is needed with a proposal that now follows the infinitesimal approach in that only details of already known things will be investigated, which delights the granting agency tremendously as the promise to obtain "positive" results is almost guaranteed, and you have your grant funded. Congratulations. Thus after months or even years since you originally started having the ideas for the research —depending on when the agency has the calls for applications— the project can commence. Hopefully you will still be interested in it!

Here I furnish an excellent illustration of what is in the granting agencies (well, their reviewers) minds. We submitted a grant application to an organisation that promised (exact words in their website): "[Name of organization] will support truly **innovative** and therefore **risky** research". We received the reviewers' comments to our proposal, stating that: "Proposal is **very original**, looking at fluctuations [….]; **Risky**, it has potential […]". So far so good, one would think, as we fulfilled the two major factors the institution was seeking in applications (those in bold characters above). The continuation of their text was the following: "… Risky: it has potential but we don't know whether it will work"; the grant was rejected. Now you see. So much for innovation and risk. They want projects almost guaranteed to "work", but my project was real research, therefore indeed we did not know whether it would work! That's precisely the fun of doing research, exploring the unknown! It seems that in our days one has to explore the already known. Anyway, further words on this topic, after this example, are superfluous. I will just reproduce, because of its witty and prophetic character, the following excerpt about how to wreck science from Leo Szilard's "The Mark Gable Foundation", that, despite being finally published in 1961, was written in 1948, so this is a really foretelling passage; it is remarkable that even in those early times the conditions to come were so apparent already.

How to retard science? Set up a grant programme (by L. Szilard)

"Well," I said, "I think that shouldn't be very difficult. As a matter of fact, I think it would be quite easy. You could set up a foundation, with an annual endowment of thirty million dollars. Research workers in need of funds could apply for grants, if they could make out a convincing case. Have ten committees, each composed of twelve scientists, appointed to pass on these applications. Take the most active scientists out of the laboratory and make them members of these committees. And the very best men in the field should be appointed as chairmen at salaries of fifty thousand dollars each. Also have about twenty prizes of one hundred thousand dollars each for the best scientific papers of the year. This is just about all you would have to do. Your lawyers could easily prepare a charter for the foundation...."

"I think you had better explain to Mr. Gable why this foundation would in fact retard the progress of sciences", said a bespectacled young man sitting at the far end of the table, whose name I didn't get at the time of introduction.

"It should be obvious," I said. "First of all, the best scientists would be removed from their laboratories and kept busy on committees passing on applications for funds. Secondly, the scientific workers in need of funds would concentrate on problems which were considered promising and were pretty certain to lead to publishable results. For a few years there might be a great increase in scientific output; but by going after the obvious, pretty soon science would dry out. Science would become something like a parlor game. Some things would be considered interesting, others not. There would be fashions. Those who followed fashion would get grants. Those who wouldn't would not, and pretty soon they would learn to follow the fashion, too."

3.2 Academic Capitalism

In individuals, insanity is rare, but in groups, parties, nations and epochs, it is the rule

F. Nietzsche, Beyond Good and Evil (1886)

The current conditions were already anticipated early in the past century. The physicist Leo Szilard's tale shown above in the previous section was an example of accurate forecasting. Politicians saw it coming too. Dwight Eisenhower, for instance, warned in 1961 about the future of universities, saying that the university was "historically the fountainhead of free ideas and scientific discovery", but even in those times it was becoming evident that money and not discovery had become the principal target. A contract had become "a substitute for intellectual curiosity". One can thus perceive the tendency: scientific research turning into an industrial enterprise. Therefore funding agencies invest money in projects that will surely produce results, and preferably of lucrative nature, that is, commercializable results.

I have witnessed the increasing corporatization of academia. For one thing, I found somewhat surprising the language used in our annual departmental retreats — these are routine reunions where scientists discuss the state of the department or the institution— that to me sounded very much like corporative talk: strategic planning, management, specific milestones to be attained next year, and other terms that suggested to me that my institution was in its course to become a corporation. It is not without interest that the word 'milestones' has become commonplace in today's academia; the term makes sense in companies and industries that have specific things to achieve, but talking about true scholarly work, the only thing to accomplish is to perform your research day after day (there were times, in my annual internal reviews or in progress reports, that in the section 'milestones expected next year' I just wrote 'to continue performing my project about this or that'). Surprising was too that many things discussed revolved around how to make more collaborations, larger research alliances, we even had administrators who came to teach us about managing and related stuff. I thought, perhaps naively, that in these retreats we would be discussing basic research done in our programme, ideas and proposals for future investigations, but somehow these were discussed only in passing, and mostly in the bar while sipping beer with the colleagues.

Many have advanced words of caution about turning academic institutions into business. In some views, this "academic capitalism" is contributing to research fraud and mishandling of cases where some research has turned into tragedy. One of those cases occurred in the field of surgery, where a certain surgeon working at the Karolinska Institute, who apparently had a revolutionary new treatment for regenerative medicine, ended up being accused of research misconduct and unethically performing experimental surgeries; seven of the eight patients who received one of his synthetic transplants died. On another gloomy note, there is the case of the suicide of Professor Stefan Grimm. At age 51, he took his life by asphyxiation. A victim of the corporatization of academia and of the new standards to assess scientists, Grimm, a biologist at Imperial College, was threatened to be laid off because his research was not expensive enough, being placed on performance review. His institution, like many others around the globe, required the scholar to possess a certain minimum amount of annual funds (200,000 £/year was expected) derived from his grants. Grimm had difficulties obtaining such minimal grant income, even though he seemingly had enough for his research, but I guess that does not count. In his own words, that appeared in an e-mail that was sent one month after his death describing his final thoughts ahead of his death: "This is not a university anymore but a business with very few up in the hierarchy, like our formidable duo, profiteering and the rest of us are milked for money, be it professors for their grant income or students…". The email was sent from Grimm's account and apparently set to deliver on a delay timer. Conceivably, he wrote it with the suicidal idea in mind, considering one sentence there: "This leads to a interesting spin to the old saying 'publish or perish'. Here it is 'publish and perish'". The whole email, contained in the text "Publish and perish at Imperial College London: the death of Stefan Grimm", can be found in several places, including http://www.dcscience.net/2014/12/01/publish-and-perish-at-imperial-college-london-the-death-of-stefan-grimm/. Following these events, Imperial College London started to 'review procedures'.

What are some consequences of this state of affairs? Besides the new criteria to evaluate scientists, that will be the focus of the next section, just take a cursory look at some discoveries made in the past. There was a time, perhaps from the end of the 19th century to the middle of the 20th, when fundamental new paradigms appeared in science: Darwin's theory of the origin of the species, the development of thermodynamics, quantum physics, the birth of molecular biology, relativity, electromagnetism. Revolution in science leads to new paradigms. Do we see any new paradigm developed these days? It is not without interest that in spite of the tremendous technological arsenal at the scientists' disposal these days, revolutionary paradigms and fundamental new discoveries are scarce, if not null, at least compared with those times aforementioned, the times of Maxwell, Darwin, Bohr and other scholars who dared ask new questions and investigate different routes, essentially following their personal interest, their vocation. But we are not cultivating any longer curiosity, creativity, vocation. Science as exploration is rapidly dwindling; it has been declining for many decades, a decline that perhaps heralds what science will become in the near future. Very few today venture into real innovative theories or exploring very novel avenues. I had grant applications rejected because the enthusiasm of the referees was "dampened by the exploratory nature of the study", in their own words. Now, if one is in a field that does not necessitate much funding, say, a theoretical domain like pure mathematics or some theoretical physics, perchance one can make excursions into really revolutionary ideas. But if one depends on obtaining substantial funding from agencies for laboratory work, this is just not advisable... Unless one has a close relative or excellent friend of very high rank in that agency. Looks like the scientific research projects of today are driven more by business considerations or practicalities than by curiosity (e.g., the typical project given to students for their dissertations are those that will almost for sure yield good "positive" results so that the degree —Ph.D. or Master— will be awarded); and, perhaps due to the immense technological advances, many times I have had the feeling that we collect data not because we want to answer particular burning questions, rather it is because we can collect those data, because we are able to perform those experiments due to the methodology available; in this manner, some researchers become welded to technology.

An additional outcome of the new situation in academia that is preventing the advance of scientific revolutions is that it is better for you as a scholar to be extremely specialised, to become an authority in one specific topic rather than being a polymath like those scientists of old. To quote William Bialek: "Some of the giants of classical physics—Helmholtz, Maxwell, and Rayleigh, to name a few — routinely crossed borders among disciplines"; in my opinion, the wider the vision a researcher has, the more probable that a new revolutionary idea will materialise in the scholar's brain. If one looks at the advertisements for academic jobs, only a few will require the candidates to have a widespread vision of whatever the field is, but the immense majority prefer candidates that are extremely conversant in one of very few particular areas. Hence we are all specialising more and more, but I am afraid, because, as Nicholas Butler put it: "An expert is one who knows more and more about less and less until he knows absolutely everything about nothing."

Some could think that the reason why there are no recent revolutions in science is because almost all possible revolutionary concepts and theories have been discovered, there being nothing really new to be uncovered, e.g. John Horgan's "The End Of Science: Facing The Limits Of Knowledge In The Twilight Of The Scientific Age". I am not sure about this. If one asks me today whether I think there is no new paradigm to be found, it would be like asking Galileo whether he thought electromagnetism or quantum theory would be part of the future. I just do not know. We do not have enough imagination or knowledge to assert that science will end because of shortage of novel theoretical concepts (not of phenomena, because these are countless and will keep us busy for aeons, but here we talk about conceptual revolutions). Even Aristotle said, more than 2000 years ago, that "for almost everything has been found out, although sometimes they are not put together"; it is interesting that there have been characters in different eras who thought that almost all had been mastered. My intuition is that there are still revolutionary concepts to be revealed, that is all I can say.

In view of this situation derived from the new academic capitalism, some initiatives are taking place. One is the Bratislava Declaration of Young Researchers, presented in 2016, that besides calling on the European Commission (EC) to recognise the special role that young researchers play, censures the "economically oriented, impact-focused bureaucratic system" that hinders creativity, curiosity, innovation, in short, what science is all about. It is of note that this declaration is put forward and supported by young fellows, early career researchers, which perhaps presages that a change in attitude is in the horizon. One can only hope, because the existing manner of doing things in the scientific enterprise is having disastrous consequences when it comes to evaluating research and appraising individuals, our next topic.

3.3   Bibliometrica, or the Final Judgement of the Scientist

The art of being wise is the art of knowing what to overlook

William James, *The Principles of Psychology* (1890)

Another initiative that endeavours to advise how to guide the evaluation of researchers is the Leiden manifesto [1]. As a consequence of the rapidly developing corporate culture in academia, the standards to evaluate researchers are changing too. As it is common knowledge among all scientists —but not to most of the lay audience out there who still believe that scientists are evaluated in terms of their acumen— it is numbers that count, and more especially number of grants (that is, moneys) and publications is the standard for performance appraisal of scientists today. It is of note too that there are no points awarded for personality features that contribute to the making of a good scientist, like motivation and curiosity. But this is understandable, for, how can one measure motivation, curiosity and other characteristics? No, to make things easy, measurable features are needed so that, conveniently, a number can be attached and thus save the minutes or hours review panels would need to (fairly) assess the academic's achievements.

The authors of the Leiden manifesto propose ten principles to make fair the evaluation of research. They realise, as the immense majority of scientists do too (which is funny to think about: we all realise the situation, and most do not agree with it, but still here it is and not much is being done to alter it, thus the importance of these initiatives) that nowadays data are used to govern and dictate the course of science. Research evaluations that were once personalised are now routine and reliant on metrics, like number of publications, number of grants and amount of money, values of impact factors, number of trainees in the laboratory... The general term with which every scholar had to become conversant in current times is bibliometrics: the quantification of written works; and, associated with it, today we find a host of bizarre indices that attempt to further quantify scientific endeavours. These measures, while of very practical use in libraries and information sciences, are ineffective in their more modern application to be used as a proxy to judge quality of scientific research. There are so many editorials and comments that have appeared in major science journals and other venues (just try a search using 'impact factor nonsense' and you will find over twelve million results, some of them editorials) on the senseless application of these quantifications that render unnecessary the insistence on the concern that scientists have on the topic.

As with other features that have emerged in the business of science, it may not be possible to know exactly what bright mind, what enlightened person was the originator of the notion that arbitrary quantifications of papers, funds, seminars and other scientific activities could be fair proxies to evaluate research. These days, the individual scientist as well as the institution and scientific journals tend to be judged by a host of indices that bear names such as impact factors, h-index, f-index etc. Metrics that are thought to measure what, in reality, should never be measured: impact, creativity and productivity in research. As such, grant applicants to several agencies are required to list not only some of these ethereal indices but also to record more figures, like the number of citations of their publications. Recently, I was applying for a grant and had to find out the number of citations to some of my papers; I found there are several search engines that provide these figures but, to my surprise, they are all different! Sometimes the differences were large, for example 82 and 120 citations were given to one of my papers in two of those search engines, and 259, 233 and 37(!?) citations to another paper in three sites. What numbers to write down, then? Obviously, let us choose the engine that gives the highest citation magnitudes... As you can see, more arbitrariness is introduced in the process of evaluation of individuals, as if there was not enough already. I do not know whether there is a single scientist in the world who thinks worthy the effort to compile all these digits and indices that are totally arbitrary and many times miscalculated —because the information needed to compute the indices cannot be accurately retrieved (we have seen how my citations differed considerably depending on the search)— but I am sure administrators and politicians see nothing wrong with these imaginary numbers, even more imaginary than the number i (which is defined as $\sqrt{-1}$).

The creators of these indices are known. For instance, E. Garfield co-invented the impact factor (Cartoon IV), but not to be used to rank researchers, which he openly declared already in 1998; here are his own words: "The source of much anxiety about Journal Impact Factors comes from their misuse in evaluating individuals, e.g. during the Habilitation process. In many countries in Europe, I have found that in order to shortcut the work of looking up actual (real) citation counts for investigators the journal impact factor is used as a surrogate to estimate the count. I have always warned against this use. There is wide variation from article to article within a single journal" [2]. For those interested in details, a short review on the impact factor from its computation to its limitations can be found in Ref. [3]; equally informative is the document elaborated by R. Adler, J. Ewing and P. Taylor on citation statistics [4].

When the impact factor started to be used, in its origins around the 1960 s, it was in fact useful to librarians and others in similar domains as a way of comparing scientific journals and to manage journal stocks. But then, why was it extended to judge scientists' performance? The main reason, once again, can be found in the time issue. It takes much less time to evaluate someone if one looks at one number, rather than thoroughly scrutinising the individual's deeds. The problem is that it is a crude statistics that is totally useless to compare different journals across disciplines, or the individuals in different disciplines, or the real impact of a publication or a discovery —for it occurs quite often the impact is felt several years in the future, whereas the journal impact factor is estimated every two years. It is as though we were only concerned with our productivity in the very recent past. Therefore do not be surprised if you are asked to list, in a grant application, what you think your best papers are in the past decade or so, as though what you did 15 years ago has now become worthless.

In a very brief interview Nobel laureate Martin L. Chalfie expresses his thoughts on one of these indices, the most known of all, the infamous impact factor ("What do you think of impact factors? https://www.youtube.com/watch?v=sCAsAKgNPjs), and if you watch it you will hear him saying "I can categorically say I hate impact factors"; then, for about two and half minutes, he explains what the original purpose of the factor really was. Suffice to say it was never meant to judge institutions or individuals, but, at the time of the writing of this text, the aberration continues… Today, almost every scientist agrees that a journal impact factor should never be used to evaluate the individual's research; I would even say that the entire scientific community agrees with the words of Adler et al. "The sole reliance on citation data provides at best an incomplete and often shallow understanding of research". And yet, we still practise it! The ludicrousness of this situation with these surreal quantifications is perfectly exposed in an article by H. K. Schutte and J. G. Svec in *Folia Phoniatrica et Logopaedica*, which, because it cited all articles published in that journal over two years, more than doubled the impact factor of that journal, from 0. 655 in 2005 to 1.44 in 2007. As the authors openly state in the paper, talking about their intention to write the article with the sole purpose of increasing the journal's impact factor: "While we realize that this initiative is absurd, we feel it adequately reflects the current absurd scientific situation in some countries".

The Comment written by the biochemist G. A. Petsko in the journal *Genome Biology* exemplifies the nature of the issue, and it is so humorous that it is given below in its entirety.

Having an impact (factor)

Gregory A. Petsko

Address: Rosenstiel Basic Medical Sciences Research Center, Brandeis University, Waltham, MA 02454-9110, USA.

Email: petsko@brandeis.edu

Published: 29 July 2008

Genome Biology 2008, 9:107 (https://doi.org/10.1186/gb-2008-9-7-107)

The electronic version of this article is the complete one and can be found online at http://genomebiology.com/2008/9/7/107

The time: Some time in the not-too-distant future.

The place: The entrance to The Pearly Gates. There are fluffy clouds everywhere. In the center is a podium with an enormous open book. A tall figure in white robes with white hair and beard stands at the podium. Approaching is a thin, middle-aged man with glasses and a bewildered expression. He is the soul of a recently deceased genome biologist.

GB: My gosh is this...? Are you...? Am I really...?

St Peter: Yes, I'm St Peter. And yes, this is where souls such as yours enter heaven.

GB: Wow. I mean, I didn't expect to live forever, but still, this is something of a shock. (Pauses.) OK, I guess I can live with it. Uh, I mean...

St Peter: I know.

GB: Well, at least I'm here. I'm not thrilled to be dead, but it's a relief to know I'm going to heaven.

St Peter: I'm afraid it's not that simple. We have to check.

GB: Check what?

St Peter: Your life history. (He leafs through the enormous book.) It's all here, you know.

GB: I'm sure it is. I can imagine you guys keep records that make PubMed seem like a stack of index cards. I'm a little surprised you don't use something more up-to-date, though.

St Peter: If you mean a personal computer, no— we don't. After all, they were invented elsewhere.

GB: You mean on earth?

St Peter: No, somewhere a lot warmer. (He stops at a page.) Here you are.

GB: Hey, I'm not worried. I was a good scientist, a good citizen, a good family man, I think, too. I never…

St Peter: Yes, yes, I'm sure, but you see, none of that matters. The only thing that matters is your IF.

GB: IF?

St Peter: Your impact factor. That's all we use now. If your IF is above 10, then you enter here. If it's lower, well…

GB: My impact factor? What the hell - oops, sorry - is that?

St Peter: It's something we borrowed from you science chaps on earth. Oh, we used to do it the hard way: send a fledgling angel down to check on your deeds; look at how your life affected your friends and family, consider your intentions versus your actions. All that sort of thing. It was tedious and required huge numbers of new angels, who have become somewhat scarce since free-market capitalism became all the rage down there. Then we noticed that you scientists never bothered to do anything like that. If you had to evaluate someone, all you did was look at this number called the impact factor. So we did the same thing. Now when anyone comes here, all we do is look up their number.

GB: A single number? Are you nuts? You can't sum up someone's whole life in a single number!

St Peter: You do. At least, you sum up their career that way, when you decide if they've published in the best journals or done the best work. It's how you work out who gets promoted and who's a star and who gets funded and…

GB: Yes, but it's a terrible idea! We should never have done it. It ruined European science in a matter of a few years, and then it spread to Australia, China and Japan, and finally to Canada and the US; and before too long, science was totally controlled by unimaginative bureaucrats who just used that number for everything. It was a disaster!

St Peter: That's not what St Garfield thinks.

GB: St who?

St Peter: St Eugene Garfield, PhD. He invented citation analysis, remember? He thought using the IF was a great idea - really, a logical extension of his own work creating the Citation Index. So we set it up: for example, I see here that you contributed regularly to several local charities.

GB: Of course. They do good work. I never did it because I thought it would get me into heaven, but…

St Peter: Just as well, because it won't. Local charities, you know. Small impact factor. Doesn't really add much to your total. Besides, how bad could the idea be? Why, the journal Genome Biology advertises its impact factor right at the top of their website. Didn't you use to write a column for them? (He looks at the ledger again.) Oh my, I see that won't add much to your total either.

GB: But that's all ridiculous! It's the whole problem I was trying to explain to you. That's like saying that a paper only has significant impact if it's published in Nature, Science, or Cell. Once you do that, then the impact factor of where you publish becomes a surrogate for the use of your own judgment. No one bothers to read anyone's papers when they're up for a fellowship or being considered for a job or for a promotion or having their grant proposal evaluated; all you do is look to see how many papers they've published in high-impact journals. No one considers whether the work was better suited to a more specialized journal or a journal where other work that puts it in context was published previously; no one considers whether those handful of high impact-factor journals have the best referees or whether they in fact may have a disproportionate number of incorrect papers because of the pressure to publish there. And look, over reliance on one stupid number gave a small bunch of editors enormous power over the careers of people who, for the most part, they never met or heard speak, and whose body of work they never read. It was probably the worst idea since General Custer thought he could surround the whole Sioux Nation with a couple of hundred troops.

St Peter: Ah, yes. St Sitting Bull still talks about that.

GB: Huh? (Shakes himself.) Look, once the impact factor dominated scientific judgments, creative people were doomed. Bureaucrats didn't need to know anything or have any wisdom; all they had to do was rely on arbitrary numbers. And now you're telling me you're doing that to determine who gets into heaven?

St Peter: Yes; it's a lot simpler. It doesn't matter if you were kind or tried hard or did good work or were pious or modest or generous. The only thing that matters is how big an impact we calculate you had.

GB: But that's just wrong! Look, maybe I could talk to the people who thought up that idea and pushed for its use. If I can just get in for a minute…

St Peter: Oh, they're not here. (He waves his hand and an image appears on a cloud. It shows a huge pit of boiling sulfur and brimstone. In it, up almost to their necks, are a bunch of men in business suits.) As you can see, they're in a warmer climate.

GB: Well, at least, that seems fair somehow. Wait a minute - is that George W Bush?

St Peter: Yes.

GB: But his impact factor should have been huge.

St Peter: Oh, the absolute value was off the charts. But we do take the sign into consideration…

GB: Then why is he only in brimstone up to his knees?

St Peter: Oh. He's standing on Dick Cheney's shoulders. (The image vanishes.) Now let's get back to you…

GB: But don't you see, the idea that you can determine someone's impact in the future from where they publish today is totally absurd. On that basis, God would have an impact factor of zero. I mean, He did his best work a long time ago; it has never been repeated by anyone; and all His ideas were published in a book, not in a peer-reviewed journal!

St Peter: Very funny. Go to hell.

CARTOON IV

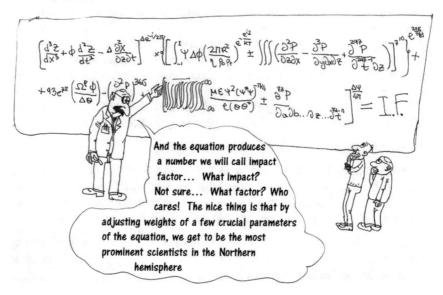

The origin of the concept that these quantifiers have value ranking individuals probably belongs to the scientific and administrative community as a whole; it is like asking who invented the cart pulled by oxen, or the initiator of taxes; it is a collective phenomenon, albeit a disgraceful one. But it could be of interest to at least know how it happened so that it is not repeated —learn history so that we do not stumble twice on the same stone. The conceivable scenario leading to the spread of this practice stems from the lack of time that scholars have to devote themselves to evaluating the works of peers and institutions, and from the lack of knowledge that bureaucrats and administrators have about science. While the former factor is to be expected in a world where there are more and more scientists every day —and thus many more things to judge— the latter should not be a factor at all, because bureaucrats should not become involved in science more than the minimum required, but their involvement has become deeper and today much research is dictated by governments, corporations and policymakers. At the same time, it is a current trend in our societies that almost everything has to be measured, counted, quantified, ranked, and, make no mistake, it is all about profit. Science, being a facet of society, became infected with this same predicament.

The unfortunate consequence of impact factors and the several indices used to quantify research is, once again, that risk-averse research is preferred [5] and that innovative research is stalled. The h-index, for example, is a measure of the scholar's productivity and citation impact of the publications, and because it is based on the most cited papers and the number of citations that they have received in other publications, then it stands to reason that if you desire a fat index your work has to be cited by many, and fast. Therefore, you should work in a popular area of research, where legions of researchers are working and thus the probability someone will come across and read your publications and cite them are higher than if you were to do research in a very innovative field where almost nobody works on.

In addition, these arbitrary quantifications foster poor quality research because it encourages quantity over quality (Cartoon V). Recall the previous comments on the irreproducibility issue and the decline effect; good quality research needs time, not only to think but also to perform the experiments and analyse the results. An outcome of the contemporary pressure to publish is the proliferation of science journals. Apparently journals are profitable considering how many are around and, besides, they continue to appear at an astounding pace; e.g., during the past 3 or 4 years I did not have one month go by without being asked to submit a paper to a new journal. To quantify a bit, let's look at the number of neuroscience journals in 2000 and in 2017: 104 rose to 154. We are flooded with papers, scientists are drowning in a sea of data. It is barely possible to keep track of the research being done even in a very specialised field. But, does this proliferation represent a real growth of knowledge? Here one should recall the previous comments on Kuhn's "development-by-accumulation" of more and more details about accepted theories and facts. Already in 1965, D. J. de Solla Price said: "I am tempted to conclude that a very large fraction of the alleged 35,000 journals now current must be reckoned as merely a distant background noise, and as very far from central or strategic in any of the knitted strips from which the cloth of science is woven" [6].

Institutions —and individuals alike— become obsessed with rankings. But scientific research is not an Olympiad, although, amusingly, one finds things like The International Mathematical Olympiad, the World Championship Mathematics Competition for High School students… But this is another matter! There could be a fastest calculating mind, but there is no "best scientist in the world"; each one of us, hauling our limitations after us, tries to advance knowledge and the comprehension of nature that, sometimes in the near and other times in the very distant future, will benefit mankind, while at the same time satisfying our curiosity for natural phenomena.

Undoubtedly, the present state of affairs is too unsatisfactory for some organizations which are breaking free of the bibliometric mentality. One of these, the University Medical Centre Utrecht, emphasises the personal, subjective perspective that candidates applying for jobs have of themselves by asking them to write an essay about their past research and future plans and a brief description of what — subjectively, yes, but at least this being more real subjectivity than that of indices and similar quantifications— they consider their greatest achievements. It is not without interest that the scholar's subjective feeling about his/her published research is a better indicator of the impact than any factor associated with the journal where the work is published. In the same manner that "publication in a journal doesn't magically transform data from conjecture into fact" (words by the geneticist Vincent Lynch), publication in a high impact factor journal doesn't make the work more valid. Some scientists have mentioned the point that each researcher has an accurate feeling about what work has had, or will have, more impact, and indeed my own experience is an example of this. I have published in journals with very high impact factor, like Nature, and in others of lower ranking; thus, one would expect the Nature paper to have more impact than others. Yet, I sort of know what my most influential work has been. It appeared in a lower ranked journal, a decent one but nowhere near Nature's impact factor. One measure of impact is number of citations, so when I looked at those, the data corroborated my feeling: the Nature paper was cited by 71 other papers, whereas the one I consider my most significant creation was cited by 263 (at the time of this writing), being my most cited paper. So much for impact factors!

Let us acknowledge too that (very) few words have been written in support of such indices and factors. One reason advanced in favour of using indices is that quantitative performance metrics may provide more objectivity when it comes to research evaluation. However, this "objectivity" falls apart when one considers how these numbers are computed —recall the *Folia Phoniatrica et Logopaedica* story aforementioned— as journals know what to publish to increase their impact factors. For example, because reviews tend to be more cited, the number of review papers published has significantly increased in some journals. Because the factor is a measure of the frequency with which the average article in a journal has been cited in a particular year —the formula being simple: the numerator is the number of current year citations and the denominator is the total number of articles published in that journal during the previous two years— then some editors receiving manuscripts for publication in their journal ask authors to cite papers from that journal (learn more about these tweakings in Ref. [7], "The top-ten in journal impact factor manipulation"). As well, it has been known that some journals negotiate the factor, for instance, by convincing the agency in charge of calculating these factors (today it is Thomson Reuters) to "adjust" certain number of items published, such as removing conference abstracts from the denominator —recall the formula abovementioned, decreasing the denominator will increase the quotient. Another known case is that of Current Biology whose impact factor increased by 40% after acquisition by Elsevier in 2001 (see details of all these, to some extent amusing, incidents in Ref. [8] "Deep impact: unintended consequences of journal rank"). In sum, that there is not much objectivity left in this business: "While numbers appear to be "objective," their objectivity can be illusory" [4].

There is another initiative that intends to halt the practice of correlating the journal impact factor to the merits of a specific scientist's contributions, the San Francisco Declaration on Research Assessment (DORA), and you can sign the declaration visiting https://sfdora.org/. As one can see, these types of initiatives that endeavour to fix the system are emerging little by little.

The paradox is that, as was already aforementioned, these metrics normally do not measure time spent in the laboratory, hours used pondering about experiments and conjectures, which precisely is what a scientist must do (I apologise for the many times I insist on this point, but seems to me that the oblivion of this is a key factor to the somewhat unpleasant, or strange, situation in research). As an illustration, here are the six points used in the scientists' annual internal review in one institution:

publications;
grant funding;
national and international recognition;
mentorship, research training and teaching activities;
collaborative activities;
institute/university administration.

Do you see any point that gives you a score for spending hours in the laboratory? For doing experiments? You can obtain excellent marks in any of those six points without stepping into your laboratory more than twice a month. This is how you do it, in six easy steps: stay at home or in your office writing grant after grant; obtain a few grants that provide you enough funds to hire several technicians and trainees; put them to work in the laboratory doing experiments; allow them, graciously, to write the papers ('so that' —you tell them— 'you learn how the writing of scientific material is done'); publish those papers and use them to obtain more funds and become renown; start the loop at the previous step 2. I have even known PIs who had their graduate students write the projects for grant applications, which would still be a bit fair if the trainee's name was added as an applicant of the grant but normally only PIs are allowed to figure as applicants.

An interview with Sydney Brenner, Nobel laureate in 2002, exposes the situation in a very eloquent fashion:

"[*Interviewer*]: It is alarming that so many Nobel Prize recipients have lamented that they would never have survived this current academic environment. What are the implications of this on the discovery of future scientific paradigm shifts and scientific inquiry in general? I asked Professor Brenner [*SB*] to elaborate.

SB: He wouldn't have survived [*he is talking about Fred Sanger who won twice the Nobel prize*]. Even God wouldn't get a grant today because somebody on the committee would say, oh those were very interesting experiments (creating the universe), but they've never been repeated. And then someone else would say, yes and he did it a long time ago, what's he done recently? And a third would say, to top it all he published it all in an un-refereed journal (The Bible). So you know we now have these performance criteria, which I think are just ridiculous in many ways. But of course this money has to be apportioned, and our administrators love having numbers like impact factors or scores."

And later he continues expressing his opinion that, as one can appreciate, is identical to mine and to that of scores of scientists:

> "**SB**: That's the difficulty. In order to do science you have to have it supported. The supporters now, the bureaucrats of science, do not wish to take any risks. So in order to get it supported, they want to know from the start that it will work. This means you have to have preliminary information, which means that you are bound to follow the straight and narrow. There's no exploration any more except in a very few places"

The full interview "How Academia and Publishing are Destroying Scientific Innovation: A Conversation with Sydney Brenner", highly recommended that young fellows fully read, can be found in several websites.

CARTOON V

3.4 The Fascination with Numbers

The obsession with quantification of scientific productivity has gone beyond the appraisal of the individual's productivity and journal prominence to invade many other aspects. It affects the writing of research protocols and grant applications. For instance, it is common that scientists have to declare the percentage of time they will devote to a project when they write the grant application. It goes like this, after writing the main text of the grant application describing the project to be done, a section will be found where one has to write the percentage effort of the applicant, say, 50% effort means I will devote, as a principal investigator, half of my time to this project if it is funded. 100% means I will devote all my time to it. But what do these percentages really mean, when talking about science? They represent an extremely rough and most likely incorrect approximation to what in reality will occur. That metric is supposed to be based on a 40-h week (8 h of work per day), but research is not a 9 AM to 5 PM job, the typical of many office jobs in North America. I have said in other parts of the book that scientists normally are working almost all time, even if asleep. Only technicians can have a 9 to 5 job, but for sure not PIs, or trainees. For them, it is close to a 24 h job. Besides, and going again to the most crucial point of doing real research, we can only barely anticipate what will be found and thus what we will do. I may plan to work full time in a project involving some experiments that I know very well how to execute, but if it turns out the results obtained point to another direction, to other types of experiments in which I know little, then I will not participate as much as planned. In essence, the requirement to write down these percentages is pointless, and I would say that, once the project is funded and starts, nobody cares about those numbers, the precise percentages will vanish in the sea of research.

Another very common futile exercise in quantification emerges when we write ethics protocols for either animal or cognitive and clinical projects —and occasionally in grant applications too— where we are required to specify the number of humans or other animals we foresee to use in the experiments. Since this protocol is based on a research project, and once again stressing the fact that in true research we don't know what will be found, then the estimation of the numbers of subjects that may be used will be very approximate and only a range can be provided, say, I plan to use between 20 and 40. The problem is that sometimes a specific figure is requested because the protocol will be stored online and the software does not allow a range, rather it requires a number, thus one must declare that exactly 25 subjects will be used in the study. Fine, the software is happy, nobody complains. Now wait one year; in the renewal of the protocol the following year, when I submit the progress report of the research done during the year, I will be asked to write the number of experimental subjects that have been used and if it is not very close to that figure of 25, then I will have to spend time giving the reasons why either fewer or more individuals (or animals) were used. I realise this may be a minor thing in terms of time waste—just to explain in two or three sentences the reason will not be a great burden— but the problem is that there is a multitude of minor, and others not so minor, things that all together preclude my focusing on research. In grant applications some agencies require the specification of research subjects too, and related to this is the sample size estimate and the so called power analysis, which is used to determine the optimal sample size for a study such that an adequate power to detect statistical significance is achieved. This is a standard procedure in large clinical studies, perhaps clinical trials, as it is crucial to have an estimate of the number of patients needed in a study. This is reasonable, but in the end, if we are performing true research— at the risk of being too repetitive I will stress again what this means: unknown results—the estimates are nothing but extremely rough approximations to the numbers that will be used to complete the study. Some scientists, more especially those in clinical fields, take these power estimates too seriously as though the numbers were more factual than the results that will be obtained during the course of the research. It is conceivable that if the project follows the aforesaid infinitesimal approach to research, then a power estimate may be close to what will develop, but otherwise, in other projects more adventurous the results of the investigation at each step cannot be known with any certainty and therefore the estimates will be nothing but science fiction. Again, then, this power analysis is one more load for the researcher, one more calculation that may not be needed at all, but that has to be done when writing a proposal lest I receive it back with criticisms for not being too accurate. And once more, this may be another minor thing, but just take a look at the many minor matters that call my attention during a normal, scientific day. I present you a day in the life of a scientist:

I enter my office in the morning and upon turning on the computer I find a few emails that require my immediate attention. One may be asking me to review a paper for a journal; another is from a prospective student or postdoc asking me to take him/ her in our laboratory; there could be an email from the purchasing department indicating something went wrong with an item we were buying; and then another message directs my attention to an ethics protocol that is about to expire and needs to be renewed. I thus have to spend some time handling all these messages, answering the potential trainee (as I always try to give my response to those asking these type of things as fast as I can, just out of respect), looking at the paper and see if I have the knowledge to be a referee or not, then I will have to download the form for the protocol renewal and start working on it. The morning thus almost went by, and presently one of my laboratory colleagues comes and tells me that an inspection of our laboratory will occur in a few days and a form has to be filled, or could be telling me that another experiment needs to be done to address this or that question (at least at this point I get to talk a little about science!), for which an addendum to our ethics protocol has to be written and has to be approved by the ethics panel. So the morning passed and lunch time comes, but instead of lunch I will go for a coffee across the street to a café and will bring with me a paper or two to read, in this manner I will be away from telephones (I did not have cell phone!) and computers— that is, emails— and will be able to concentrate on those papers. Upon my return to the laboratory or to my office, I find out we have to buy more reagents for an experiment but since one of them is a controlled substance— perhaps a drug used in animal experiments— a form has to be filled out and submitted to the appropriate governmental agency for permission to buy a certain amount of the compound, amount that has to be justified with numbers of all sorts, like number of animals we will use in the experiments and the dosage per animal and how many times the dose will be given a day or per week etc. In other words, I have to come up with more imaginary numbers because, once again, this is research and we just don't know. Some effort, as can be appreciated, will have to be invested filling out this form. And since we plan to buy more reagents, I have to be sure we have enough money in our operating grants, so I have to look online to the state of my grants, and as many times occurs, I notice that something is amiss with the amounts included as expenses in one grant. So I have to investigate, I may call or visit or send an email to my grant accountant and ask about that potential mistake reported in my expenses. The afternoon thus proceeds. Presently, around 3 PM, I look at my watch and have to decide whether or not to do the experiment I was planning, an electrophysiological recording in vitro using cell cultures, knowing that it takes about one hour to prepare the electrophysiology set-up for the recordings, perhaps 3 or 4 h to run the experiment, and about thirty minutes cleaning the set-up after all is done. If I start the experiment, then, I will leave the laboratory around 8 or 9 PM. So perchance I will do it tomorrow. But tomorrow I must spend time searching online for funding agencies to resubmit the previously rejected grant, and start re-formatting the proposal according to the particular agency… Anyhow, now you have an idea how most of my time goes, and thanks that I am not a very important person within my institution I do not have to spend too much time with administrative or semi-scientific meetings, because for other colleagues of mine high in the hierarchy, the time they spend meeting in panels or other similar stuff is substantial, say, once or even twice per day. Now, some readers may be asking, when do I, or they, do the experiments, then? When do I analyse the data I collected? When do I reflect to interpret the results? Well, there is night time, after all the day has 24 h! One of my colleagues told me once that the majority of his research thinking— reading papers, contemplating experiments — was done at night. And with this observation, I end my account of a normal day in my "scientific" life.

3.5 The Rise of the Megagroups

The contemporary statu quo seems to be missing, or ignoring, a fundamental factor in the progress of science: the most important factor in scientific advancement is the individual scientist. Individual curiosity is the main driving force of science, but individual creativity and the lone scientist are species close to extinction (e.g. Ref. [9] "The demise of the lone author"), as the institutions —including governments, universities, hospitals— favour the formation of large corporations, the implementation of huge projects, not all being strictly research; for instance, there are organisations that provide vast sums of money for infrastructure—say, erecting a new building— but offer little, or nothing at all, for research operating purposes. I can provide many examples but will mention just one: once upon a time our

institute received and spent millions in the purchase of a new magnetoencephalography machine, in the preparation of a special room for it and in related
infrastructure costs, but we were not given funds to hire personnel, technicians that
could operate the sophisticated device. We were forced to use our own operating
grant money to hire a technician. The following cartoon (Cartoon VI) was inspired
by these events.

CARTOON VI

Here, children, we see the remnants of another casualty of those times. The
poor wretch passed away just at the moment he was about to press the
"submit" button for his 32nd individual grant application that year, trying to
obtain funds to hire a technician who could help him unpack the $19 million
pentaphoton confocal hypermicroscope that remains still inside those cages
you see here, for which, incidentally, he had no problem finding money to
purchase it thanks to a multi-billion mega-group super-consortium grant; the
same granting agency that gave him those funds built the NeuroInfonomics
research centre you see across the street through that window, but since they
had money only for infrastructure and nothing to hire personnel to move into
the building, it has remained empty until now. Rumour has it that ghosts of
scientists have been seen in the halls.

It is true that many projects executed these days require large groups, for example to sequence genes in the human genome project, or to do large-scale computer simulations in projects like the Blue Brain, or in massive clinical trials involving dozens of hospitals and thousands of patients. The main feature in these enterprises is that very few get to think, and the majority are data-generating personnel. But these are particular, large scale projects. The tendency, though, is the formation of large groups even in not-so-large-scale ventures, as if more heads together meant more ideas, better innovation. In fact, I would say the opposite is true: there is no linear relationship indicating that scaling up the number of scientists working on a common project scales up the creativity and insight. Here is an example illustrating that point, taken from our experience interacting with the military on traumatic brain injury projects: the Canadian armed forces set up a small research project with very limited funds and obtained results they could interpret, while the US army, thanks to their massive funding, performed a very extensive study that involved many groups but, as they revealed to their Canadian counterparts, they had substantial difficulties making sense of the enormous amounts of data gathered. Sometimes small is clearer. Some readers may have heard the terms Big Data or Big Science, these are commonplace today in science; the former refers to extremely large data sets that have to be computationally analysed to extract meaning of all those vast datasets, revealing patterns, correlations, etc. The latter denotes the aforesaid research that is very expensive, involving many teams of scientists. Big, therefore, is a trend today in research.

Have we forgotten that many of the main scientific revolutions came about with almost no funding and from scientists working basically on their own? I do not think scientists have forgotten, but politicians and administrators for sure... Or perhaps they never knew (Alain Trautmann, one of the nine European scientists that wrote the open letter "They have chosen ignorance", published in The Guardian, www.theguardian.com/science/occams-corner/2014/oct/09/they-have-chosen-ignorance-open-letter, said politicians have an "abysmal ignorance" of what research really is). Einstein did not need many grants to ponder about relativity; Gödel discovered his incompleteness theorems without the help of a vast group or large sums of money; Fleming found penicillin out of his personal curiosity while looking at petri dishes; Darwin just needed someone to finance his trip on the Beagle; Ramon y Cajal spent some of his own money to fund his research on the structure of the nervous system. Money, while it helps, is not an absolute requirement to make revolutionary discoveries. Accentuating once again the point: the individual is the main driving force in science.

These comments are not meant to criticise the formation of research groups. In fact, these have some advantages, among those is that trainees and junior investigators become introduced to new colleagues with whom collaborations may be established. Another benefit of belonging to a large group is that one can increase the number of publications rapidly: if each researcher in a group of 20 writes one paper and includes everybody in the author line, then each of the components of the group will end up with 20 papers per year; this reflects, and enhances, the highly competitive nature of academia (see Chap. 6). The predicament is that the formation of big, oversized groups, consortia and infrastructure teams are greatly favoured by funding agencies over the individual scientist, there seems to be no balance. It occurs similarly, as aforementioned a few paragraphs above, in the case of grants awarded to short-term, commercializable research and to long-term, not-so-practical investigations, the absolute emphasis is in the former. The enlargement of research groups has another adverse consequence in that large laboratories or groups of laboratories can obtain lots of data that will serve as pilot data to apply for more grants and obtain more money, hence the large will become larger, while the small laboratory will continue dwindling. It can be clearly seen today the concentration of resources in laboratories that are becoming bigger, sometimes engulfing (as I have witnessed in my career) others not too wealthy.

To sum up, the increasing corporatization of academic institutions is posing a severe limit in the kinds of research questions allowed, which in addition to the funding policies, is favouring hackneyed research and creates an unfair and to a great extent illogical evaluation of scholars. Besides the preference for short-term, utilitarian research, there are other related aspects that emerge from the current system that will be treated in the next sections; one aspect is that it urges to publish papers in "high impact" journals and the associated peer review concern; another important matter derived from corporatization is the notion of intellectual property; and finally, the emergence of journals as private, lucrative enterprises. But before commenting on these aspects, and considering the trend of quantification of research, we should come up with solutions to the current situation before the time comes when our careers as scientists are summarised by a single number. Let us thus suggest a few recommendations to junior researchers about how to navigate the corporate scientific system.

Possible solutions— how to write the imperfect grant for an imperfect system

Thus, having considered all the previous comments on the current funding mechanisms and scientific standards, this section offers two lines of action to remedy the shortage of funds for your research (please remember this is addressed to novices, professional academics know these tricks very well!). The first is the 21st century recipe for a successful grant application and, in general, for a wealthy laboratory, whereas the second advice is for those who want to pursue their own research interests although these may not have an utilitarian facet.

Every scientist knows the tricks to write a grant application with high probabilities of being funded, there are no secrets. Start by choosing a project that is fashionable; for instance, in biomedicine, choose a project that will have a great impact on healthcare. In other fields pick whatever is clear that has immediate utilitarian features. Most importantly, whatever the project is, be sure it will yield "positive" results, those that were aforementioned in previous sections. Another fundamental point: the referees reviewing proposals for the funding agencies seem to have forgotten that science is an exploration, because many grants are rejected on account of the proposal being too exploratory, thus, I can almost guarantee that if you write a truly exploratory project, no matter how creative and wonderful it may be, it will not be funded. Hence, be sure to make clear your proposal does not sound like an exploration of a phenomenon, even though in reality it could be —in fact it should be, if it is a true research proposal; you just have to conceal this exploratory nature (see the section 'The lottery of grants' for more on how to use the proper "vocabulary" when writing grants). What funding agencies like to see is a clear hypothesis at the start of the proposal, and the project has to be based on this hypothesis (or hypotheses, you may have more than one, but not too many), recall the comments on Chap. 2 regarding the absolute preference for hypothesis-driven proposals rather than question-driven projects. At this point one may think that writing a grant application is sort of writing a science fiction text, since so much has to be hidden from the "expert referees"—that is, if you actually plan to do genuine, authentic research: exploratory, creative, risky, possibly question-driven. Well, perhaps that is the reason why I am also writing now a science fiction novel, after so much grant writing during my career. And on this topic, the style of writing, write the text clearly and when possible use simple words, including technicalities of course (after all it is a research text) but be sure it can be understood by non-specialists in your field, because most granting agencies use several reviewers and some will not have the precise expertise. Now we are arriving to the issue of peer-review, which will be commented upon in the next section; the peer review matter is mentioned now because in the final analysis whether your grant is funded or not is, mostly, a totally subjective decision that depends on the biased opinions of a few reviewers. Hence it does not matter you think you have written the perfect grant, for the 'perfect grant' does not exist because the reviewers are not perfect, rather are prejudiced, many times ignorant and may have their own agendas. More on this later. Nevertheless, a project written with those characteristics aforesaid has good chances that sooner or later will be funded. One advantage is that in these days many funding agencies have specific programmes, and therefore it is advisable to send the grant to one of these if the project fits the requirements of the call. There are as well agencies that prefer to fund research in a specific problem disregarding the popularity of the theme, so it is not always a matter of choosing a stylish area; for example, there are institutions devoted to strange, uncommon diseases. But send your proposal to as many agencies as possible, because, as any scientist knows, having a grant funded is a lottery— see the section below on the lottery of grants — hence, the more numbers you buy, the greater the chances.

Now, if you are one like me who wants to pursue his own research interest, there is still hope to maintain, if not a wealthy laboratory, at least a relatively well-funded house where you can investigate what you really want to address. The trick is to write the prose in such a manner that a very practical feature of the research becomes evident in your proposal. Or, choose a granting agency that specialises in funding more holistic, pure research; yes, there exist this type of agencies, although there are very few. Another trick is to have one project that is very trendy or has immediate applicability, obtain a grant with this research and use those moneys to cover some costs in another, more theoretical project that almost nobody will fund. A warning here: one must be honest always; if you are given money to do such and such project, it is fair to the agency that funded your research that you carry out that project, I would not advise to use those capitals entirely in other projects, this is not fair, remember to give to each his/her own. But some funds can always be diverted to those other more holistic, perhaps 'unfundable', projects. This may depend on the funding organization, some are very strict and require that their funds be used strictly in the described project. That is fine; a paid technician or trainee can always find time to work in your favourite project.

The preceding recommendations were just a few specific workarounds to increase the probability of obtaining funds. But the global solution to the basic problem of the distribution of funds depends more on the agencies and policies than on the individual. Considering that administrators and politicians in charge of funding bodies will be disinclined— and I am afraid they will still be for many years to come —to fund creative and risky research, one possible solution that may be advanced is that while they can keep the largest amounts of moneys for the more "secure" research that will yield results in the short term, at least they could create specific programmes for the start of those more novel research proposals that lack preliminary results, because, as it happens, if you have a revolutionary idea for a project, the chances that there is solid preliminary evidence, or pilot results as they are called, are tiny— this is why grants are awarded normally to projects with such a substantial magnitude of preliminary observations that basically guarantees that positive results will be obtained. These special programmes could provide money for pilot projects of short duration so that evidence can be obtained regarding their plausibility. It was aforementioned in the section on closed loops (Sect. 1.1) the existence of pilot project programmes, but, as indicated there, there is still need for preliminary evidence; what I am proposing here is an authentic pilot project programme: no evidence whatsoever, other than the purely theoretical or logical, would be asked.

Possible solutions—Moving beyond metrics

Once we accept the fact that it is virtually unfeasible to provide an accurate estimate of the impact and efforts of a researcher, for scholars are almost never off the job even at night during sleep— how many times have I dreamt about the questions I was facing in my daily research, but perhaps one of the most famous discoveries made while dreaming was that of the cyclic structure of benzene by F.A. Kekule —then we are forced to go beyond metrics, to move beyond the quantification of individuals and organisations. Cartoon VII is based on a real occurrence that casts more evidence in the obsession of organisations to quantify research output; once they implemented the 'research subject registration', where we were asked to record and keep count of the number of individuals who volunteered for neuroimaging recordings (those subjects who helped us record with magnetoencephalography, functional magnetic resonance or electroencephalography their brain activity in neurocognitive experiments). Of course, it is customary in research to document experiments done and results obtained in things like laboratory notebooks, therefore looking at my notebooks one can have an idea of everything I did. But this is, apparently, not enough, because notebooks are not bureaucratic utensils; subject registration was a formal, administrative manner to keep track of the numbers of experiments performed, and, at least in its inception, was a pain to do: for example, imagine the scene, we had to register adult volunteers going through the hospital registration procedures and, being a children's hospital, it was an interesting sight to see adults standing next to children and the corresponding confusion of the desk attendant who was supposed to "admit" the adult in the hospital. It became easier later on.

CARTOON VII

One immediate help anybody can execute is to support and sign, wherever possible, the aforesaid initiatives like DORA (https://sfdora.org/), the Leiden Manifesto or the Bratislava Declaration. But to me, the first question that should be answered is whether or not there is the absolute and inevitable necessity to evaluate research in this strict fashion. Great discoveries were made in old times when the appraisal of scholars was much less severe as it is today, or basically nonexistent. Of course those were very different times with very few academics, whereas today there is a myriad, as commented upon in the chapter below about competition in science. Because there is not enough money for everybody and therefore it has to be allocated, some manner of judgement as to who obtains the funds has to exist. But what if the judgement were minimal, tending to infinitesimal? Let us envision a possible scenario.

Imagine for a moment that we dispose of the standard metrics like h, impact, and other factors. Continue envisaging, if it is not too much to ask, that while we take a look at the publications of the individual submitting a grant proposal that— for the purpose of this example —we are reviewing for a funding agency, we do not count the number of publications and furthermore we do not even pay attention to the journals where his/her publications have appeared. What do we have left in the CV, aside from other professional aspects like teaching and training of fellows? There is still the list of publications— the full list and not only those published in the past 5 or 10 years, as some agencies demand and that I find unfair because I want to know what the applicant was doing 15 or 25 years ago to have a better picture of the applicant's achievements and interests —that should give us a good idea as to whether or not the applicant can accomplish the research proposed in the grant. This fact, coupled with the importance and novelty of the proposed project, should result in us forming an educated guess, or opinion, as to the priority of this grant against another that we will review next —for, as it happens, referees in panels have to review several grant proposals. In the final analysis, then, we would be judging whether the applicant has asked a relevant research question and whether the team has the means to answer it, without much consideration of quantifiers like factors, number of publications or moneys the applicant may have.

If instead of a funding panel we are talking about the standard annual evaluation of a scholar in an institution, then the question comes up again: is it needed an annual (or bi-annual, or n-annual) review of each academic? For sure some measures have to be implemented such that scholars do work. The only thing I can say is that if I were the person in charge of reviewing the performance of my colleagues in the institution I would probably do as I always did with the selection of the graduate students I took: I briefly, very fleetingly, looked at their CV, and then talked to them. By direct conversation and exchange of thoughts, I had already a good idea what the aspiring student could or could not achieve. Judging thus— and to be honest —I committed some, but very few, errors. But the regular evaluation of scholars would be even simpler, as I already know what they have accomplish over the past few years and of their interest and other personal dispositions— in prospective students one never knows the person in front asking you to take him/her into the laboratory, so there is an element of risk. I also would have in mind that research is a slow process and that, as many times mentioned in this text, requires time to do it well; I would not be blinded thus by number of papers published annually or laboratory size or personnel hired by the scientist. Therefore, in my institutional review of academics, I would not have used any factor, any counting of funds, publications or employees. Problem is, I would have had to spend more time in glancing over the publications and achievements, plus talking to the scientist in a manner that would give me an idea as to whether the scholar is working on his/her questions about natural phenomena or whether he/she is in a permanent, self-imposed, vacation. In the case of the former, the academic under my review would have my words of encouragement for the time period to come before the next inevitable internal (or external) review. If the latter, well, the director, chairman or president of the organisation would have the last word!

References

1. D. Hicks et al., Bibliometrics: the Leiden Manifesto for research metrics. Nature **520**, 429–431 (2015). https://doi.org/10.1038/520429a
2. E. Garfield, The Impact Factor and Using It Correctly. www.garfield.library.upenn.edu/papers/derunfallchirurg_v101(6)p413y1998english.html
3. P. Dong et al., The impact factor revisited. Biomedical Digital Libraries **2**, 7 (2005). https://doi.org/10.1186/1742-5581-2-7)
4. R. Adler, J. Ewing, P. Taylor, Citation statistics. Statistical Science **24**(1), 1–14 (2009). https://doi.org/10.1214/09-STS285
5. P. Stephan, R. Veugelers, J. Wang, Blinkered by bibliometrics. Nature **544**, 411–412 (2017). https://doi.org/10.1038/544411a
6. D. J. de Solla Price (1965) Networks of scientific papers. *Science*, 149(3683), 510–515. https://doi.org/10.1126/science.149.3683.510
7. M.E. Falagas, V.G. Alexiou, The top-ten in journal impact factor manipulation. Archivum Immunologiae et Therapiae Experimentalis **56**, 223 (2008). https://doi.org/10.1007/s00005-008-0024-5
8. B. Brembs, K. Button, M. Munafò, Deep impact: unintended consequences of journal rank. Frontiers in Human Neuroscience **7**, 291 (2013). https://doi.org/10.3389/fnhum.2013.00291
9. M. Greene, The demise of the lone author. Nature **450**, 1165 (2007). https://doi.org/10.1038/4501165a

Chapter 4
Money Matters—Pay or Perish

In part as a consequence of the corporate culture in academia is the fact that whatever a scholar does or produces belongs to the institution. This is another aspect that the non-initiated may find surprising; for example, my ideas that ended up in patent applications were bought by my institution for the sum of… 1 (one) dollar (Cartoon VIII). And if someone does not believe it, here is part of the text of the document where I sold my idea for such an impressive sum of money:

> AND WHEREAS, [name of institution] Herein after referred to as the "Assignee", whose full post office address is […] desires to acquire the Assignor's entire right, title and interest in and to the Invention and the Application; NOW THEREFORE, in consideration of **the sum of One Dollar ($1.00)** and other good and valuable consideration, the receipt and sufficiency of which is hereby acknowledged, the Assignor confirms it has, as of June 9, 2015 sold, assigned, transferred and set over, and for greater certainty, does hereby sell, assign, transfer and set over to the Assignee, its successors, assigns, or legal representatives…

Thus the concept of Intellectual Property (IP) entered academia. It refers to the protection of the creations of the intellect that may have a commercial value. Now, it is fair that through IP laws those who have achieved innovations have rights for exploiting and benefiting from their creations, this is reasonable. But, as many times occurs, the fair notion of IP has been stretched in academic institutions resulting in, perhaps unexpected, outcomes affecting the individuals working for these organizations. For, as it happens, it is not only the scientific idea that belongs to the organization, but other things too including choreography or artistic works (!), as can be inspected in the following paragraphs taken from the Intellectual Property document of one of the institutions where I worked: "Intellectual Property includes but is not limited to technology, lab notebooks, technical information, formulae, computer software and hardware, drawings, graphics, designs, concepts, ideas, apparatus, processes, research tools (including without limitation, biological materials and other tangible research property and devices), and all original literary, dramatic, musical, and artistic works (including without limitation, books, architectural works, choreographic works, and cinematographic works), computer programs, all print, multimedia electronic and audiovisual parent/patient information materials, manuals, program packages, and educational material".

© Springer Nature Switzerland AG 2019
J. L. Perez Velazquez, *The Rise of the Scientist-Bureaucrat*,
https://doi.org/10.1007/978-3-030-12326-0_4

In short, it seems that the organization owns all I do, think, dream and produce. This volume would be their property too, according to the previous text, but it is written after my time there. Now, it belongs, in part, to the publisher!

CARTOON VIII

It has been my experience when talking to lay people about these matters that some pecuniary facts directly affecting the individual in the world of science are perceived as very surprising. Among these financial aspects, besides the afore-mentioned proprietary steps that organizations take with the researcher's ideas, one finds the curious event that even though we scientists publish lots of things, we do not make almost any money; just the contrary: we must pay to publish.

4.1 The Price of Knowledge

One peculiarity of the scientific business that is somewhat surprising, at least that has been my experience conversing with people, is that scientists publish a good deal of papers in journals or conferences proceedings, chapters in books, and books to a less extent, and yet they do not make any money (perchance a bit in case of books); rather it is the other way around, we must pay in order to publish. And since according to the old scientific aphorism it is "publish or perish", then it is concluded that we have to pay to survive: pay or perish. Most of the journals will charge publication fees for figures in colour and sometimes for the text itself like the so-called open-access journals that have fixed publication fees that depend on the format of the paper (they could be research papers, reviews, opinion papers etc.). As an illustration, the article processing charges —the price to publish a research report in open access journals— range from about 350 to 2500 euros. More classical journals that are not open access sometimes do not have a publication fee, but if you include figures in colour, then be prepared to pay anything from 150 to more than 1000 euros, depending on the journal. Folks not in the field may be surprised to know that papers need to have colourful figures; I will let you know that figures in the papers are used to visualise the results of the experiments and many times black and white will not suffice, colour is needed so that the graphs, plots, and sketches are comprehensible. Take a look at almost any research article in a journal like, say, Nature or The European Journal of Neuroscience, and observe the figures... You will see why colour is necessary. However, because very few hard copies of journals are printed (today it is all in the net and papers circulate in PDF formats), in reality it makes little sense to pay today for colour figures, so one can opt to have them printed in black and white —makes little difference as very few read hard copies of journals these days.

The aforesaid open access journals offer free access to the public to any paper published there. This is a great idea, it is as it should be: the general public should have free access to scientific reports if only because part of the funds for research are coming through the taxes we all pay; taxpayers thus have the right to know what people in white coats are doing with their moneys. Other journals, the classical journals which are the majority, have a restricted access and will charge a fee if one wants to look at or download a paper. As we can see, all is business. And open access too, as we saw that these journals charge a large fee to publish there.

Several thousand of researchers have expressed their discontent with the restriction to the spread of knowledge, and advocate for sharing scientific results; they went so far as to sign up to a boycott of journals that restrict free sharing as part of a campaign that has been called the "academic spring". It is not only unfair but —it is my opinion— ridiculous that if you are interested in looking at one of my publications you must spend valuable minutes of your life registering with a journal (this is not always the case) and paying the fee to download the paper. Scientists publish, among other (imposed by the system) reasons that have been commented upon in this volume, to share the knowledge: this is and should be the principal reason for publication... Although as we have seen there are other administrative considerations that impel one to publish in this time and age. The executive director of Research Libraries UK, David Prosser, said that "To be made effective, scholarly information has to be made as widely available as possible". There is no point in papers being stack in bookshelves behind the walls of academic institutions where almost nobody will reach them. In current times, with the incredible advance in internet, search engines and communications, any person in the globe —with a computer or the means to have access to one or similar device— should be able to look at any publication whatsoever, without being charged fees, regardless of the moderate price the fee may be. The increasing interest in these matters can be appreciated by the input of major organisations, like the European Commission that published a report in 2017 on "Providing researchers with the skills and competencies they need to practise Open Science".

This current situation in scientific publishing is unsatisfactory from many viewpoints and it has several unfair consequences that affect scientists. For example, in an effort to foster the spread of knowledge, some funding agencies, like the Welcome Trust, imposes academics that obtain funds from them to publish in open online journals. But, as aforesaid, open access publication fees are very expensive, which forces scholars now to include in their budgets extensive amounts of money to pay fees. Considering a normal experimental laboratory in the biological sciences that publish around 5 to 8 papers a year, this can easily amount to about 8000 euros per year in fees, but many grants that are awarded offer very limited funds, therefore it is not a trivial matter to find money to pay for these types of papers. And the situation is more unfair in other academic fields that, because these are not experimental, have much more restricted funds. It is then not surprising that once or twice I had to discuss with the journal chief editors the fact that a philosopher had written, or wanted to write, a paper for a special issue I was editing but he could not afford to pay the publication fee as he did not have grants. Normally, you do not need much money to do philosophy —even though there are experimental philosophy laboratories where expensive experiments may be carried out— therefore these academics will have to use money from their own pockets to pay in open access journals. Scientists in developing countries have a scarcity of funds which prevents them from publishing in certain journals of "good standing"—to be fair, let us mention that many journals offer discounts and waivers to scientists in developing countries. This is one aspect that should change as soon as possible, so that all academics, regardless of field or country, have the opportunity to publish where they wish. But of course, to criticise is easy, to propose solutions is another, more difficult, matter. Nevertheless, the Possible Solutions section contains some ideas to at least start fixing this situation.

And some solutions may stem from the aforementioned Academic Spring. This is the name given to a campaign that, as of today, about 17,000 researchers have signed up in order to boycott journals that restrict free sharing of the published reports. This initiative could revolutionise the spread of knowledge, but because of the desire of profit among publishers of scientific journals it may take a while to fully develop into something fair not only for the academic community but also for mankind in general, as everybody would have access to any research done around the globe. All these movements are ultimately due to the concerns about the rising cost of publishing in academic journals and retrieving papers from them. If you take a look at the website 'The Cost of Knowledge' (http://thecostofknowledge.com/) you will have an idea how many people are disturbed by the situation: more than 17,000 academics have registered their protest. And, apparently, all this started with a blogpost by a certain mathematician, Tim Gowers, who was unhappy with the statu quo. One can see that things can emerge from an individual action!

Several publishers are becoming aware of the interest among scholars and institutions that scientific literature should be widely available, and thus some open access repositories have been created like ResearchGate (a very large science social network where one can find anything from papers to jobs and collaborators etc.), arXiv (containing preprints, for physics mostly), bioRxiv, (preprint server for biological sciences), SharedIt (with the purpose of legally sharing contents of papers); and there are several other initiatives with similar purpose in mind, like CORE (COnnecting REpositories), a service provided by the Knowledge Media Institute to aggregate all open access content distributed across different systems such as repositories and open access journals.

4.2   For the Love of the Art

It could be of interest to the general public to realise that publishing is not the only undertaking researchers in academia perform "free of charge", without receiving remuneration, so let me give you a short list of unpaid activities. Giving counsel is one activity, perhaps the most common all academics perform, free of charge. This may be somewhat striking too, because employees in other trades with which lay people are more acquainted will charge fees for almost every minute of their services. So, as a consultant, you may charge a —many times very substantial— fee to those requesting your guidance. Yet, the advice of academic scientists is almost never paid —only in special circumstances, and of course the situation of those working in industry and corporations is different, for after all here you work for money. During my career I met with entrepreneurs, clinicians, the military, members of other academic institutions, and provided what was, to be sure, my perhaps inconsequential piece of advice on whatever topic my very limited skills were needed. I have a vague recollection of someone in a company asking me for guidance and saying that of course I will be paid for that —what I best recall is my surprise at the natural way with which this fellow mentioned this matter of paying me for the advice, probably because for him it was an ordinary event to be paid for this type of things— and I must admit I did not know what to say (no, I did not request payment in the end). The list of unpaid work continues with editorial services. I am in the editorial board of a few journals and have never received a cent; perhaps the chief editor may receive some salary, but normally those in the editorial board do it for free, for the love of the art. As well, when we act as referees reviewing papers in that infamous peer review activity discussed below, nothing is paid.

I have been working for two institutions during the past several years, and one of them, the university, did not provide me with a salary. This is also a matter of astonishment to some of my uninitiated friends. How can one work for a place that does not pay you? My situation is in fact quite common in research. I was hired by a hospital that paid me fully but I acquired an affiliation to a university, which is what normally occurs to those not working directly for the university, like clinicians in hospitals. It is true that in some cases both organisations share the remuneration, but not always. Thus I became a Professor in the university and taught a graduate course there, again, for the love of the art.

Why do we do these things? some readers may now be wondering. There are two possible answers, one very general that applies to everybody, the other more peculiar for a certain population of scholars. In short, the general answer is that we do those free-of-charge tasks because they improve our CVs. To be editor of a journal, or a participant in panels, or having a professorial appointment in a university looks good on your CV. And one can benefit from it too, for to be affiliated with a university means that you can receive students in your laboratory— that is, workforce for the laboratory— who may proceed towards a graduate degree. In exchange for these favours from the university, you write your affiliation in your papers, which is good for the university in our ultra-competitive world, plus you do some little tasks for them —at least in my case they never asked me to perform so many things as others, who received salary from them, were forced to do. For instance, I was never asked to teach. I went a bit further, it is true, and taught a graduate course for a few years —in fact, perhaps paradoxically, I had to persuade them to let me teach the course while at the same time not asking for compensation, it was almost as difficult to prepare the forms and materials to go through the review for approval of my proposed course as preparing the course itself!— and was the organiser of a shorter version of the course later, but I wanted to do it because I liked it: I designed this course based on a topic of major interest to me —consciousness and self-awareness— because there is nothing like teaching to improve your comprehension of a subject. Hence, I did not mind the extra, unpaid work, because it was not work after all, but my own gratification; so you see, I did it on purely selfish grounds —and this is the second, more restricted answer: some of us do these things because we truly enjoy them. Besides, it is fully acknowledged that in academic research one is supposed to disseminate knowledge, to help others understand nature, therefore those advices we provide or materials we teach, that we scholars give for free is, in reality, part of our jobs. It is understandable how alien some of these things may sound to those working for corporations or other service-oriented business; which gives me again the opportunity to warn the student: doing research in the industry or in academia are two very different things.

Possible Solutions

One solution already in practice to solve the problem of the limited spread of knowledge is the open access publications. I must admit that I am an advocate of open access journals, which may have transpired through the previous paragraphs. Yet, some problems exist, and especially the unfair situation derived from the relatively large publication fees that preclude academics in certain fields (e.g. humanities), or countries, to publish in open-access journals. Waiving fees for academics in non-experimental fields or in developing countries is an immediate solution that, as said above, is implemented by some journals. Due to the nature of the world today, especially in its economic aspects, publishing without fees may not be feasible. As long as there is money around, and as long as there are many journals, someone will have to pay, either the librarians and readers or the authors of scientific papers. It can be perhaps envisaged that reducing the number of journals may help decrease the fees —related to this "solution" of decreasing journal numbers please see the Possible Solutions paragraphs 'The Reviewers' Office and the pruning of journals' in Chap. 5— but still the existing journals would have to hire personnel and spend money in the publishing, hence fees will always be there. I do not know whether the current fees charged for texts or colour figures are reasonable or too high (they will never be too low, of that one can be certain). Could the national governments assist the publishers? This may not be possible either, considering that politicians are not investing much in basic research, so asking them for more funds to "spread knowledge" may not be in their typical 4-year perspective.

Chapter 5
The Tragicomedy of Peer Review—The Publication Game and the Lottery of Grants

> *Lack of progress in science is never so much due to any scarcity of factual information as it is to the fixed mindsets of scientists themselves*
>
> F. R. Schram, 1992

Scientists, whereas they tend to be a smart bunch who perform very careful experiments and reach reasonable conclusions after interpreting the experimental results, are human after all, thus limited in their acumen and prone to make mistakes, just like any other person in any other trade. Hence, the standard procedure in scientific research has been, since time immemorial, that the experiments, results, and interpretations obtained by one are scrutinised by others, peers expert in the field. This is termed peer-review. But it is not only used to review papers sent for publication, it is done as well in grant applications, and in other things like ethics protocols.

There has been some sort of review process since times past, or at least a cursory evaluation of the research presented, but the modern concept of peer review started to be applied around the early 19th century, when the Royal Society started to seek referees' reports; or, it could have been as early as the start of the 1700s, when, apparently, the Royal Society of Edinburgh was publishing some collections of peer-reviewed medical articles. Reference 1 provides some historical facts about the beginnings of peer-review. There have been other types of "reviews" in antiquity that I would not really qualify as peer review, like the censorship exercised by religious organizations; recall the very famous Galileo affair in the early 1600s.

5.1 The Publication Game

While the concept of reviewing papers to make sure there are no mistakes is praiseworthy, the current situation is rather futile. There are several reasons. First and foremost, the peers should have enough expertise in the field —this is obvious — but most importantly, and here we find again the topic encountered everywhere

© Springer Nature Switzerland AG 2019

J. L. Perez Velazquez, *The Rise of the Scientist-Bureaucrat*,
https://doi.org/10.1007/978-3-030-12326-0_5

in this volume, there has to be enough time to evaluate the experiments, results and conclusion the paper reports. It is the time matter, yes, one crucial facet that makes peer review a failure. The starting words of A. Csiszar's "Troubled from the start" [1] are precisely "Referees are overworked". After publishing over one hundred papers, it is my experience that the most common review is that of someone who did not invest enough time to understand and judge properly what is written in the paper. Same can be said about reviews to grant applications; many times it seemed the reviewers did not even read the narrative of the proposal.

I have many, I would say, amusing examples about the apparent lack of attention of grant reviewers, but two are my favourite that will serve to illustrate the point. Once our team wrote a grant proposal and was criticised because we did not have enough expertise in the experiments we proposed; these experiments were mostly neurophysiological recordings in animals, rodents in particular. The reviewer completely ignored to look at our biographical sketch —as it is common that with a narrative describing the project, a sort of curriculum vitae is sent along with the package. Either our reviewer did not read it or read it while falling asleep, because I had been performing rodent cerebral recordings for at least a decade, and my collaborator in the project had spent all his professional life, more than 30 years at that time, doing intracerebral recordings in animals; yet, we were deemed not to have enough expertise in in vivo neurophysiological recordings. In another occasion, we were asked to remove one part of our proposed project, involving recordings in rats… But we did not propose any! This was a re-submission of a grant application and it was in the previous submission that we had mentioned rat experiments, but in the re-submission these were omitted because the reviews to our first submission did not see the rodent experiments worthy, so we removed them in the next round. Hence, what did the referee read? Looks like the comments to the previous submission, but not the new one!

These types of criticisms that denote attention lapses are common too in reviews of papers. Every scientist can tell you how many times referees' criticisms or comments are received indicating that the authors of the paper have failed to say or address this or that, when it was in fact mentioned and addressed in the text of the manuscript (by the way, this 'manuscript' word should not be used, as nothing anymore is written by hand, yet it is a standard term in the field); countless times. The illustrations I have given represent the standard scenario in peer-review, do not think these are uncommon reviews. And the culprit, the majority of times, is lack of attention which normally is caused by haste, lack of time to do a proper review, to scrutinise in detail the paper or the project in case of grant applications.

It sometimes occurs too that the reviewer does not have enough proficiency in the subject matter of a paper and therefore mistakes are made in the review process. This is becoming more apparent in our times. The reason is that many studies today are multidisciplinary, or interdisciplinary if you prefer —now, pundits will tell you that both words do not exactly mean the same thing, but for our present purposes we can concede them to be synonymous. Among the considerations that are complicating the process to review a scientific paper or a grant application, multidisciplinarity is a relatively novel aspect that is coming up in this day and age. Some disciplines are more likely to be interdisciplinary, for instance neuroscience. Erik de Schutter has expressed the challenge in reviewing neuroscience papers in the current age, in "Reviewing multi-disciplinary papers: a challenge in neuroscience? [2]. In earlier times, papers rarely had 3 or 4 very different techniques or methods used in the experiments. However, take a look at a few papers in some common scientific journal today, like Nature or Science, and you will see 4 or 5 figures with data obtained using same number of techniques, some figures being almost impossible to understand —so compressed the data presented in those figures are — showing results in experiments done with various methods. For example, in a biological paper one may find microscopy, electrophysiology, biochemistry and molecular biology; all together, yes, in a single paper. This rarely occurred in the early 20th century of before. It is because today there is an immense arsenal of methodologies that allow us to tackle a question from diverse perspectives. We are not as limited as in old times. But this great advantage of modern science is creating a great problem too when it comes to reviewing papers and grants. In spite of the emphasis on multi-disciplinarity, most of scientists remain very specialised; a referee may understand the biochemical experiments described in an article but has no clue about the con-focal microscopy there reported, another reviewer may be an expert on microscopy but has no idea about the electrophysiological recordings that are shown as well in the paper. One may think that this should not be problematic, just invite 4 or 5 reviewers to comment on the paper, each with a different expertise that all in global encompass the data presented in the article or project. Ideally, this should be the standard: at least 4 or 5 opinions on a paper or a grant proposal. But normally one receives only 2, and extraordinarily 3 or more.

The low number of referees used to judge a paper or application is one aspect that makes peer review a sort of farce, because it is human nature that each individual has his or her own opinion, and scientists are no different. If just two opinions are asked, there is the possibility of ending up with two totally different views. How many times —again, another trip to infinity — does one receive comments from two reviewers, one saying great things about the study, the other slaughtering it. Hence, the editor has to decide whether to reject or accept the paper based on a diversity of opinions. Let me again show you the words in Csiszar's article [1] following the aforementioned sentence on overworked referees: "The problem of bias in intractable" (it is not the last time we meet Csiszar, his next words will appear shortly). The same occurs, and more often, in reviews of grant proposals, because here there is a greater degree of prejudice —since this is a competition for funds— than in reviewing articles, as the reviewer has to express his/her opinion about how worthy the project is to be funded.

Thus, along these lines, the total futility and the farcical nature of the peer-review process is exemplified in comments like these that I received once from 2 reviewers of a paper:

Reviewer 1: "this paper does not introduce sufficient novelty…"

Reviewer 2: "manuscript is very interesting and provides novel insights…

You see the problem now, both reviewers read the same paper, and yet totally different opinions! Some readers may think these diverse opinions on the same paper do not occur frequently, and that this example is an outlier of the most common reviews one obtains; not really, this example shown above is an illustration of a very common review. In fact, receiving very different views on a paper has been there since the very beginning of peer review; W. Whewell, one of the pioneers of the system, and J. W. Lubbock reviewed a paper in the early ages of peer review, around 1832, and had opposite opinions, in part due to the different general views on research of both characters [1]. We scientists wished this phenomenon was an oddity, an outlier, but no, this is, I would even say, the standard flavour of comments to a paper; ask any other scientist if you don't believe me. Mind you, perhaps in other academic disciplines the situation is different, say, in philosophy, art, or humanities in general. That I do not know, my work has been exclusively in biological related studies; but something tells me that these diverse reviews to one study occur in every discipline. In fact, one would think that it would be precisely in the humanities where reviewers could submit completely opposite opinions to the same study because of the not-so-quantitative nature of the research in these disciplines as opposed to the "hard science", experimental research of biology, physics, and the like. Well, you have now seen it. If one considers human nature, the impossibility to avoid subjective views becomes apparent.

Human nature also dictates that criticisms to a submitted paper will reflect the disposition of the reviewers; as such, pride, arrogance, prejudices, biases, aggressiveness, inflexibility and other common features of human personality become apparent in the criticisms expressed by referees. Thus, young fellow, be ready when you enter what I call the publication game to receive negative reviews because one reviewer does not want your paper published on account of the closeness and similitude to the reviewer's own research; or because another referee has felt him/herself hurt due to the fact that you did not cite his/her published work; or because a third referee feels you should not be entering the field, with a publication in a good journal, in which he or she is a world expert. I have received several comments to papers where there was almost no criticism to the experiments themselves, rather it was apparent that the reviewer was not happy we were intruding his/her field. Scientists, let me emphasise once again, regardless of their acumen, are human. And some do not have qualms letting their ferocity out. If you have been publishing regularly it is almost a certainty that you have received, at least, one aggressive review. It is true these belligerent criticisms are not the norm, but it is somewhat surprising —yes, despite human personality features— that this occurs in what is supposed to be a very rational enterprise. Editors should not allow hostile reviews, for these represent a lack of respect and common sense in the business of academia. When I am sent a paper to review, I will tell you what I see in it: the work of, at least, one person who has bothered to write the article and go through the excruciating tedium of the submission protocols; the labour of a set of people that devoted time of their lives to study a matter that was of interest to them; the will to communicate and share their results. These considerations lead one to have a respectful inclination when it comes to judging the work. Leaving one's ego aside is also a convenient practice, which makes criticism fairer and more dispassionate; whoever is really into research should have a modest ego as one understands our transient existence in a mainly indifferent universe.

Some readers may have noticed a little flaw in the logic that some referees are needed to confirm or criticise the work done by a team in a paper submitted for publication because the team may have committed mistakes. The flaw stems from the fact that the referees, being human too, can make themselves mistakes. Thus to have it perfect, perhaps other reviewers should review the original referees' comments on the study. But then again, these second-order referees may make errors too, so maybe there should be a third-order referee revising the previous ones... You can see that we are getting into the hunting grounds of infinity. One must thus acknowledge that to achieve the perfect state in this theme is unfeasible lest we reach an infinite regress, nonetheless having just some first-order referees looking at your work sometimes helps.

It has been my experience that I had ideas I wanted to investigate and, upon performing a literature search on the topic, I found that someone had done it. The following paper is about one of those experiments I always wanted to try, and I dare say, many of us would have liked to try. A study that demonstrates —as if more proof were needed— the ineffectiveness of the peer review process. Douglas P. Peters and Stephen J. Ceci studied in "Peer-review practices of psychological journals: The fate of published articles, submitted again" [3] the "peer-review process directly", as they say in the abstract which, because of its interest and amusing nature of the "experiment", I transcribe entirely:

"A growing interest in and concern about the adequacy and fairness of modern peer-review practices in publication and funding are apparent across a wide range of scientific disciplines. Although questions about reliability, accountability, reviewer bias, and competence have been raised, there has been very little direct research on these variables.

The present investigation was an attempt to study the peer-review process directly, in the natural setting of actual journal referee evaluations of submitted manuscripts. As test materials we selected 12 already published research articles by investigators from prestigious and highly productive American psychology departments, one article from each of 12 highly regarded and widely read American psychology journals with high rejection rates (80%) and nonblind refereeing practices.

With fictitious names and institutions substituted for the original ones (e.g., Tri-Valley Center for Human Potential), the altered manuscripts were formally resubmitted to the journals that had originally refereed and published them 18–32 months earlier. Of the sample of 38 editors and reviewers, only three (8%) detected the resubmissions. This result allowed nine of the 12 articles to continue through the review process to receive an actual evaluation: eight of the nine were rejected. Sixteen of the 18 referees (89%) recommended against publication and the editors concurred. The grounds for rejection were in many cases described as 'serious methodological flaws.' "A number of possible interpretations of these data are reviewed and evaluated."

Upon reading that abstract of the paper summarising the results, one can see the futile nature of the peer review process that is being played in the circus of the publication game. No more words are needed.

If you think I am the only one expressing a somewhat unfavourable notion of peer review practice, look at the words of the Nobel laureate Sydney Brenner stating his ideas on the subject: "But I don't believe in peer review because I think it's very distorted and as I've said, it's simply a regression to the mean. I think peer review is hindering science. In fact, I think it has become a completely corrupt system. It's corrupt in many ways, in that scientists and academics have handed over to the editors of these journals the ability to make judgment on science and scientists. There are universities in America, and I've heard from many committees, that we won't consider people's publications in low impact factor journals".

Brenner's last words above point to a common notion among lay people and, as well, among some scholars, that whatever is published must be true, or close to the truth, and even more if the study has appeared in a so-called high impact factor journal. Recall Vincent Lynch's words presented in a section above, "Publication in a journal doesn't magically transform data from conjecture into fact". I, along with many others, do not think that publication in a high impact factor magazine makes any study closer to eternal truth, regardless of what the referees said about the study during the review process. Healthy scepticism is a natural trait among academics, all studies are equally debatable, more even now in our era of impact factors that in my opinion are causing more inconvenience than services, and to look down on studies published in not-so-important journals is prejudiced, to say the least. All studies, large and small, are worthy of some consideration, and whether right or wrong will not be determined by its publication in a specific journal but by time alone. As a proof, just consider what was one of the most solid edifices in Western science for 2000 years, Euclidian geometry, and how it was demoted as it was found to fail (after a long time, admittedly) to answer basic geometric questions away from the surface of the Earth. Time cures everything.

Brenner also qualified the peer review procedure as corrupt. On this topic, I would like to add that once I was informed by Springer (as I am in the editorial board of one of its journals) that fake reviewers had been spotted, hence the editor advised us, members of the editorial board, to be careful when selecting reviewers for papers submitted to our journal. Fabricated reviewers and reviews have been known to exist. It normally consists in providing a false e-mail for a genuine name of a peer in the section where one has to suggest potential peer reviewers during the manuscript submission process, in such manner that the e-mail in reality belongs to the author of the paper, and, if the editor decides to choose that suggested reviewer, then the author will receive in his/her inbox e-mail the paper for review, thus the author can write a very favourable review about his/her own article. Smart, eh?

And how about this email that a colleague of one of my former fellows received? Can this be true?

"Dear xxxxx (*we have deleted the name of our colleague*)"

How are you?

My name is Zhang, I am from China. I am a businessman, my business is about Citing for Money.

Please let me introduce the nature of this business:

As you know, the IF (Impact Factor) of a journal is very important, if a journal has a high Impact Factor, the journal will be very significant and famous. So many journals published in English ask my company to improve their Impact Factors. And the only way to improve the Impact Factors is to cite papers from the journals.

So now I invite you to cooperate into this business, the cooperation method is as follows:

The price is: 50 USD for each citation, that is, if you cite 1 paper in one of your papers, you will get 50USD, and if you cite 8 papers in one of your papers, you will get 400 USD;

Payment method: (A) bank account, USD bank account, Euro account or other bank account, if it is not USD bank account, I will send money that is Equivalent to the USD in other currencies to you;(B) moneygram company, I send money to moneygram company, and you go to branch of moneygram company in your city to get your money, if you want to know the details, please reply this email.

I will send the citation regulation and journal list to you, many journals in different research fields ask my company to improve their Impact Factors, you will find some journals that are relevant to your research;

You will cite some papers from the journals in your paper, before you cite them, you should tell me which papers you are going to cite;

As soon as the paper is accepted, you should tell me the information of the accepted paper, and I will make budgeting for you;

As soon as the paper is published online, you should tell me and send the paper link to me, and I will send money to you;

In fact, now there are many professors in different countries cooperating with my company, they cite papers from the journals, and I send money to them. If you would like to cooperate with my company in citing for money, please reply this email and we will talk about the details".

But it is not only citations that are for sale, authorship is being sold as well! In this webpage https://retractionwatch.com/2016/10/24/seven-signs-a-paper-was-for-sale/ you can read how people are selling authorship on manuscripts to be published in various respectable journals. And the more important the journal, the higher the price you have to pay to include your name on the paper. This black market for scientific papers has emerged because of the great competition in the field; next section deals with the contemporary hyper-competition in science. Some researchers are put under so much pressure to publish that forces them to go to certain extremes. Hence, new "professionals" are emerging that make their business "helping" these researchers to appease their institutions, at least in terms of number of peer-review publications. Let us mention in passing that there is a legal way to buy authorship, one that is commonly used by PIs: there are times their names appear in papers in spite of having provided a null intellectual contribution to the study, rather it is the financial contribution that guarantees that their names are added to the author line. This is one reason why you may find today scientists who publish almost 100 papers per year, the so-called hyperprolific authors ("The scientists who publish a paper every five days", a Nature Comment that can be found at https://doi.org/10.1038/d41586-018-06185-8). This after all is not that unfair, because the senior investigator normally has been in charge of raising the funds for the study. Nonetheless, this is an example of an accepted purchase of authorship.

Corruption, my friends, is present in all human activities… Except in slow wave sleep (in REM sleep there are dreams and one may have a corrupt dream!). But let us not haste to judge, prudence is advised —as always should be exercised when an opinion starts to develop in our minds due to the never ending over-elaboration of our thoughts we humans are so good at— for after all, corruption basically consists in taking advantage of the system; whoever is free from guilt can throw the first stone, because all of us at one point or another, have taken a, perhaps minuscule, advantage of the system. Of course the "advantages" taken in the aforesaid cases are of enormous dimensions.

As it stands now, and on the most crucial theme of time —that has been touched so often in this volume— the peer review system inflicts a fundamental impediment to scholars: it is an enormous waste of time. As promised, here is the third sentence of Csiszar's article on peer review [1]: "The referee system has broken down and become an obstacle to scientific progress". The system has created a common ritual among academics, submit and resubmit and re-resubmit and, so many times, $(re)^n$-submit with $n > 5$! Imagine the absurd waste of time and effort to re-write the article time after time, and this rewriting is needed because there is little consensus about a common format for papers among the (incredible number of) journals. It would really help if all journals agreed in a format so that the re-writing process is lessened, but still time would be wasted in that the re-submission to another journal implies effort in accessing the journal's website and registering, completing online forms, and other tasks. It must be considered, and fully acknowledged, that sooner or later all sound studies end up being published; so far the requirement for publication seems to be the dissipation of the scholar's energy and the waste of time re-submitting a number of times to several journals, until one finally accepts the paper. Therefore the system is not helping in this most crucial matter of time that academics enormously face, as I have stressed so repeatedly. The Possible Solutions section offers some considerations in this regard.

After the previous paragraphs suggesting that the peer-review system is somewhat a sort of fiasco, it is fair to note that some academics have found that the review they received contributed to the improvement of their papers. This is one feature that should be, in an almost perfect world —where reviews are fair, rigorous and thoughtful— a typical outcome of any review: papers should improve the quality. But, alas, it has not been my experience. In my case, I would say most of reviews have been inconsequential, and I can think only of a few papers that I consider were improved by the referees' criticisms and suggestions. But again, there are others who may have had a different experience. Nonetheless, I agree with (yes, once again) Csiszar's words in [1] that "Traditional refereeing is an antiquated form that might have been good for science in the past but it's high time to put it out of its misery" (all words cited here from reference 1 constitute the first paragraph of his article, a paragraph that, to me, concisely describes peer review today). Now, to end this section let me mention one benefit of practising peer-review. It seems to me that the trainees and early stage researches could gain experience from being referees; it is not so much to receive the criticisms and think about them, but also to be able to comment or see the flaws in studies done by others that improves one of the basic characteristics of researchers, that of precision, thoroughness. It is in fact not uncommon that in graduate school the students are asked to evaluate papers and point out the weak points or possible weaknesses of the study. Allowing trainees to perform some peer review could thus be beneficial for their training process.

5.2 The Lottery of Grants

Peer-review is not only restricted to papers. It is used in the judgement of grant applications too. Now we are talking not about the publication game, but of the grants game, as I and others call it [4]. You see, like any other aspect of life, science too is full of games that one must play according to the consented rules. And just like in the case of reviews to papers the personal traits of the reviewers play a role, so too in reviewing grants; perhaps the criticisms to potential projects to be awarded a grant are more fierce and unfair in general, as this is a direct competition for funds —the reviewer may be applying for a grant in the same competition, or in same funding agency.

To demonstrate the personal nature of criticisms to grants, much as those to papers aforementioned, here we have the following comments to a grant proposal I once submitted. Four reviewers were judging my qualifications, based on my CV, to carry out the proposed project. One evaluated me as "excellent", another as "very good", and two viewed me as "non-competitive". Four reviewers, three opinions. Regarding the project itself, one reviewer said: "the project displays a logic chain of consecutive steps", while another commented: "The main criticism of the approach is that it is poorly defined, the scientific approach and rationale is not explained". Again, looks like they read different projects. Another of my grants that had three reviewers brought the following scores (out of 10): one gave a 10, another a 7, and the third a 4, three quite different scores for the same project. And yet in another grant application I read that one referee complained that "it is not clear that he (*that is, I, the applicant*) has any experience in these models or in computerized analysis of this complexity" whereas another referee praised me saying "the investigator has a great deal of experience with dynamical systems theory and is qualified to supervise the project". To end what could be an endless sequence of opposite and very diverse opinions to the same proposal that reveal the tremendous subjectivity of the referees' comments, let me furnish another example to demonstrate how this business of writing grants is like a lottery due to the preferences of the referees and reviewing panels (a group of scientists who will go over all reviews and will take the last verdict on whether or not to fund the project described in the grant). Once I submitted an identical project to two institutions, a project that complied with the mandates of what these agencies normally fund. The project received a high ranking and was funded by one of the agencies, whereas the other gave us a low rank and did not fund it; entertainingly, one of the main reasons for this discrepancy was that one panel considered very good that I was proposing to perform experiments with which I was not too familiar, and they liked the fact that I was about to enter what was a new field for me, while the other panel viewed this as a great weakness and recommended that, first, I should become an expert in the field. You see again the nature of the problem: same proposal, two widely opposed opinions.

If you are one of inquisitive nature, you can learn about how brains work by studying the criticisms and reviews to your grants or papers. Here is one example, my favourite comment that I received to a grant application perhaps because it exemplifies main features of the review process and it illustrates too the basic functioning of the human brain. The grant proposal described a project on the role of synchrony in brain cellular activity —measured as neurophysiological electrical recordings— after traumatic brain injury; we proposed to study synchrony between brain signals in patients with brain injury. One comment in the review to our grant was, more or less: "so, what does all this have to do with chaos?" Well, nothing really, I would have answered if I had been allowed to reply. The term chaos or chaotic dynamics (some readers have probably heard of chaos theory) were never mentioned in the proposal. The reason why this comment —which is really meant as criticism, for in this game almost any comment, unless very positive, will be taken as negative by the review panel— is such an excellent illustration of the review process and human nature is because it embodies the main feature of how the brain functions in humans, and reviewers in particular. Brain processing of information is principally associative: one piece of information stored in memory leads to another related to that one, which in turn leads to another associated to those etc. We all continuously experience the associative nature of brain information processing, being the reason why some find it difficult to concentrate for extended periods of time on one specific thing. Anyhow, my project described in the grant never mentioned the word chaos, but chaos is part of the discipline known as nonlinear dynamics, and the study of synchronization can be considered part of the same field too, hence in the reviewer's mind one thing that was written — synchrony— led to the related one, chaotic dynamics; the reviewer had probably read things about nonlinear dynamics and thus both concepts were associated in his/her brain; the associative chain within his/her brain circuits started thus. This is the neurophysiological aspect of this event. But why did the reviewer have to mention it, write it in his/her report, if after all the study had nothing to do with chaotic dynamics? Because one must write something in the review report; this is how this comment/criticism illustrates a main feature of the review process: you, as reviewer/referee, must write something in the review, notwithstanding how silly or absurd the comment may be. If you do not say much on a review, the panel may think you are either ignorant, or not too smart, or both. Thus, one has to write anything that comes to mind, the more the better if you are concerned about their opinion about you. And, finally, the comment/criticism illustrates another feature of the peer-review process that was mentioned some paragraphs above: that reviewers do not read the text with attention, for it was clear that chaos was never mentioned or dealt with in my proposal. We thus have seen how a simple comment/criticism reveals three major features of human brains and the review process. The grant, by the way, was not funded.

Some of the abovementioned examples reflect personal traits of reviewers, what they consider acceptable or not acceptable curricula or professional features of the investigator. Other not-so-personal individual features influence the assessment of a project. One that was commented previously is the preferences, the mandates of the funding agency. Taking all together, the aftermath of grant writing is something that may surprise the lay audience, those not in the realm of research or those young fellows about to enter it: we sort of lie when we describe our proposals. I am not the only one saying this, as it was stated by Nobel laureates too —e.g. J. Kendrew, Nobel Prize winner in 1962: "a scientist has to cheat. If he expressed his own motivation honestly he would get no money" — and, in truth, almost every scientist knows that some lying is not only needed but acceptable. What Kendrew was talking about is that when we write grant proposals we have to conform to the mandates of the funding institution, and, as we have seen in this volume, these agencies do not fund risky and innovative projects, rather hackneyed research. Hence, we must play down in our proposal any truly novel idea we may have had and make it sound like what they normally fund, that is, describe that we plan to uncover some details about already well-known facts —what Thomas S. Kuhn would advise— investigations that are safe and relatively uninteresting (but be sure to add the qualifiers "novel, ground-breaking"!) and therefore the project will be almost guaranteed to yield positive results. If one is totally honest and describe the proposal in terms of real research, that is, that we have no idea what to expect even though we may have an educated guess of what could be uncovered, this may not be enough as the risk associated with the course to discovery may not suffice to please the funding bodies.

But do not be alarmed, dear reader, because most scientists know what an acceptable little lie is and when to stop lying, or, to put it mildly, misrepresenting reality. But, do we? The aforesaid correspondence by D. F. Horrobin, 'The grants game' [4] has something to say in this regard, subject matter that I will skip in this text. In any event, these little distortions of the real projects we have in mind that we write in proposals are not damaging to research, because when funded, scientists tend to do what they really wanted to do. Playing properly this game allows science to advance, lest we become totally entrenched in the trivial.

It can be instructive to end this peer review section with very interesting historical demonstrations that the concept of totally fair and unbiased review is nonsense. If you are conversant with some history of science, you probably know that one of the major scientific achievements, Isaac Newton's Principia, was published thanks to the efforts, and money, of Edmond Halley in 1687, who convinced the reluctant Royal Society to publish the work; even the great Newton had difficulties in his first attempts at publishing his results. Another good example, that of Ludwig

Boltzmann, one of the main developers of statistical mechanics, who had to spend his final years in heated debates defending his theories against other physicists, especially his statistical interpretation of thermodynamics, views today completely accepted (the disregard that his contemporaries had of his ideas may be a factor that sunk him into the depression that made him hang himself). In more recent times, the fundamental article describing B lymphocytes as distinct entities, one of the seminal immunology studies, was repeatedly rejected by immunology journals and appeared finally in Poultry Science —an Oxford journal— in 1956 on account of the species that were used in the study (this is not to say the journal is not important, only that it is not where one expects to find that paper). Other illustrations include the rejections of the Cerenkov radiation study and S. Hawking's black hole radiation paper. But my favourite story is that of what arguably is the top concept in modern science, the first law of thermodynamics, and the fate of its three discoverers, to wit, R. J. Mayer, J. P. Joule and H. L. F. von Helmholtz. Only Mayer was able to publish it, at least in the normal scientific fashion, after having to re-submit his paper to another journal. Joule had to "publish" it in a newspaper in Manchester where his brother worked, while Helmholtz had it rejected from journals on the basis of the study being "mere philosophy" and had to publish it as a pamphlet that he distributed among friends and relatives. Eventually, the three of them received the deserved recognition. The moral of the tale is that if you are one who tends to be vulnerable to distress if your papers are rejected or if your results are deemed inconsequential by peers, then think of the discoverers of the first principle of thermodynamics! The pedagogical story of these events is described in more detail in Muller and Weiss' book "Entropy and Energy" (Springer Verlag, 2005). And if you want more, read Juan Miguel Campanario's compilation of accounts by some scientists themselves of episodes of resistance to their new discoveries that eventually earned them a Nobel Prize in "Rejecting and resisting Nobel class discoveries: accounts by Nobel Laureates" [5].

Therefore, considering all said here about peer review, nobody will be surprised to know that there are important calls for a reconsideration of the rigour and quality of the process, with some appeals aiming at substantially changing the system. An example, The European Cooperation in Science and Technology (COST) framework has an Action called New Frontiers of Peer Review (PEERE) which "aims to improve efficiency, transparency and accountability of peer review through a trans-disciplinary, cross-sectorial collaboration." Regarding the growing frustration with the existing grant application process, Science Europe —an association of public European research funding organisations and research performing organisations— has expressed interest in trying new mechanisms in "Radical ideas required to cut research grant waste, funders told" (D. Matthews, *Times Higher Education*, 2018); you can read it and leave your opinion in reddit "Radical ideas required to cut research grant waste, funders told: New head of Science Europe says he hopes for experiments with *grant lottery system* and even a basic income for researchers", and as you see from its title, we all know this is a lottery, it is not only myself and two friends who think that way, so making it a true lottery as the Science Europe chief suggests will make little difference but may save time and effort!. Other initiatives that are starting to alter the situation are mentioned below in the Solutions section.

In spite of the deficiencies of the system in place to assess results for either publication or funding, do not rush and blame only science, for the initial rejection and incomprehension of somebody's work is present in almost all human arenas; didn't a French critic say that J. M. W. Turner's painting (a main inspiration for the emergence of impressionism) had "degenerated into lunacy"? And we know of quite a few artists that lived in poverty in their days but became prominent figures after their death. The essence of the problem with review by peers, whether in science, humanities or any other activity, lies within the skull of humans [6]. Let's see if it can be fixed a little, without the need for psychological consultation.

Possible Solutions: The Reviewers' Office and the pruning of journals

How to solve the peer-review issue is not going to be easy. Lay people not conversant with the trade may think that, well, it all starts with the individual, so the reviewer should just pay more attention and have interest in what he/she has to review so that there are no (silly) mistakes in the reviews/criticisms. This, in fact, is the simplest, clearest solution, but unfortunately I do not think it is feasible, and I hope this unfeasibility is apparent after the previous words devoted to the time issue, the administration and other chores scientists have to perform. There is just not enough time for us to review in great detail the several papers or grant applications we are asked to review every week, even every day in case of some very famous, important scholars. It would be unfair to ask them to devote energy and time to the review process and neglect the other very important matters they have to care about, some of them crucial for survival, like writing grants and papers. With the vast numbers of scientists all over the globe, each publishing several papers and writing numerous grants, it will be very improbable that reviewers will devote more time and pay closer attention to the review at hand. Neither the number of researchers nor the papers and grants they write will diminish in the near future, unless the situation changes dramatically such that we are not asked to produce a minimum number of papers per year or have a certain amount of funds. In other words, decreasing the intense competition we endure these days in science will help a bit.

Nevertheless, if the main problem is time allocation, then there is another solution: create a body of full-time reviewers, a referees' office. These characters will have no other job, no research, no writing papers or grants, their mission will be strictly to review grants and papers. The idea is not that far-fetched as it may sound; there are already some institutions that have fellows dedicated to the writing of projects for grant applications, perhaps not in all detail but still they help principal applicants with the text and other matters. Who would be the professional reviewers? Scientists, why not. There are not few researchers who are dissatisfied with the current standards of having to publish such and such number of papers every year or else… Or being demanded to obtain certain amounts of funds, some of which will cover their own salary; or other reasons. One wonders whether some of these unsatisfied scientists would not be happier in a job that does not require those pressures, only to review the papers they receive (of course the number should be reasonable, otherwise we encounter the time problem again). I have also met people who said that writing grants is their favourite aspect of science (as amazing as it sounds, yes, there are some who enjoy grant writing!), so from writing to reviewing and writing criticisms and comments to papers and proposals there may not be much of a gap. This solution would serve to eliminate one problem in peer review that was mentioned above: competition. The reviewers would not be competing with professional scientists for funds or for fame and glory in experimental results, hence they would not be too biased against the papers or grants received based on who the authors are.

But, who would pay the reviewers? Presently, the immense majority of reviewers are not remunerated. Can institutions make a business with maintaining a peer-review office? Some journals, particularly the open access journals, have very substantial fees. Could some of that be diverted to support salaries for professional reviewers? Can the industry and corporations contribute, perhaps with some incentives like considering the money they give for reviewers as donations, thereby reducing their tax? Perchance, universities and other academic centres could contribute hiring personnel for the sole purpose of reviewing, because, it is worth noting, almost half of the university employees in some countries, like the UK, are administrators (as reported by the *Times Higher Education* in 2016) —another sign of the immense bureaucratization of academic institutions— hence it would not be too much to ask to substitute some of those bureaucrats for reviewers. By the way, there is a "global community of peer reviewers": Publons. But, no surprise here, one of the missions of Publons is to turn peer review into a measurable research output, so that scholars can use their review record as evidence of their standing and influence in their field: (from their website) "Easily import, verify, and store a record of every peer review you perform and every manuscript you handle as an editor, for any journal in the world". Hence, in the final analysis, more quantifications are laid upon academics.

Finally, it seems to me one plausible solution is already out there: open access publications. This concept was treated in a previous section, and here the aspect to be emphasised of open access is that anybody can post a comment about the paper. I am not totally sure whether this applies to all open access journals but I think the majority of those have the feature of allowing readers to comment on articles. This is the real peer review, where instead of 2 or 3 pundits, potentially hundreds or thousands of individuals can provide criticisms, or praise, or whatever one thinks necessary. As for me, I can definitely say that I do not care what 2, 3 or 4 experts chosen by an editor think about my paper, I care what 400 or 4,000 —not chosen by any editor but by their interest in reading my study— think. After all, peer review should be a cooperative process between the referees and the authors (e.g. Ref. [7] "Cooperation between referees and authors increases peer review accuracy") thus the system implemented by some journals like Frontiers makes sense; here, there is a constant exchange between the referees' reviews and the authors' responses via the Frontiers website. For those interested, Frontiers in Computational Neuroscience has published a special Research Topic that includes, as of today, 21 papers on "Beyond open access: visions for public evaluation of scientific papers by post-publication peer review". The significance of the post-publication review by many, many peers, and not only 3 or 4, is accentuated in many of the articles in this Research Topic. In general, it increases transparency of research. F1000Research, an open research publishing platform for life scientists, also offers open peer review, so one can inspect the referees' comments (which normally are criticisms) to a paper.

I also have some idea about how to save the time one spends (wastes?) trying to publish a paper. Some years ago, being a bit bored with submitting and resubmitting papers to a sequence of journals and considering the aforementioned fact that any sound paper will end up being published— and our papers were sound and were finally published —I had an idea that, as it so often occurs, someone else had already had. The idea was to eliminate journals, to reduce the number of journals to a minimum, and preferably, to have only one journal. Obviously, in this case, the typical rejection based on "we think your article is more appropriate for a specialised audience and therefore not suitable for the wide readership of our journal" would not occur. In this single journal, all rigorous, good papers would be published, hence this would not change compared with what happens today, and time would be saved immensely. Of course I realise this is not feasible due to the lucrative nature of the publication business that was mentioned in another section, but it would make the scholars' lives much easier.

Around the time I was musing about this, I communicated with David Horrobin, who besides being an entrepreneur and medical researcher was an activist on the topic of peer-review, having had many publications on the matter —e.g. 'The philosophical basis of peer review and the suppression of innovation' [8] or 'Something rotten at the core of science?' [9]. He told me that long time ago, in the 1970s, he had pursued the idea of a sort of "lonely" journal with Elsevier (a major publishing company), and seemingly they created a system which ran for some time called the 'international research communications system', in which each individual would create a personalised journal based on their interest, the papers and documents they read. The idea was a bit different, though, of what I had in mind —a single journal where all researchers submit their articles— because his proposal was to have a single source where a reader is able to keep informed about everything that is going on without having to have ready access to libraries (there was no internet yet); so it was more like to have a central source of information. Nevertheless, this system did not last long, he told me, could not be maintained as it was ahead of its time… we are talking about a pre-internet era! Today, though, with the magic of internet, it would be possible to just create a megajournal where all rigorous papers are published and the readers could perform a search within that immense journal for articles of their interest. Once again, there would not be any difference in the manner the search for papers is done today: we all use search engines; very few of us still read journals, thus, in reality, where the paper ends up being published is almost irrelevant. Among the several things that have considerably changed in the scientific enterprise since the times I was a student, this is one. Technology, especially communications, has revolutionised the field in such a manner that we do not read journals anymore. I recall past times, not really so long ago, when I had to go regularly to the library of the institution or adjacent university to find papers on the subject matter in which I worked. The last time I visited the library for this purpose was, I think, about 6 or 7 years ago. Now scholars use web search to find anything of interest, for virtually everything is in the internet. I find it somewhat funny my recollections of those times when we had to mail cards —that looked very much like postcards— to colleagues around the globe asking for a reprint of their papers; these are already objects for museums of science. And more on a personal note, I miss that typical smell of the sheets of paper, the scent of the books in the library, I miss searching for the journal where the paper I want is published. Now I look at a computer monitor and smell a subtle metallic aroma coming off its back. What took us several minutes, hours or even days in those times, now I can do in a few seconds in front of my digital companion. I wonder if the new generations, post-internet, of young scientists can even imagine how slow and difficult things were in the past. Yes, things have changed, and on this topic of change, why not continue making changes to improve the system.

Who knows, perhaps in our age —which I would term the communications revolution era (coming after the cognitive, agricultural, scientific and industrial revolutions that shaped humanity, and now communications are shaping us too)— we will witness the decline of the journals. By the way, there are journals out there that claim that they will publish all papers,— regardless of the field— methodologically rigorous. This is, or at least was at the inception, what the journal PLoS One promised, endeavouring to alleviate the many inefficiencies of the contemporary system; browsing subject areas in their website feels like what I have been trying to express here (and what Horrobin had in mind): a vast journal where one can find everything. Similar ideas have been expressed by others too, for instance Bjorn Brembs and colleagues advocate an archival publication system [10]. F1000Research, a life sciences journal, claim too that they publish "without editorial bias". These platforms allow papers and readers to benefit from transparent refereeing and the inclusion of all source data. Anyhow, leaving only very few journals is a possible solution but, I am afraid, in a distant future. Horrobin died just a few months before my visit to Scotland in 2003, when I was planning to see him.

And if you still are bent on avoiding the agonising tedium of receiving peer review criticisms and submitting endlessly to a sequence of journals, you can always do like the great Gottfried W. Leibniz did: create your own journal and publish your results there! Leibniz founded, with others, *Acta Eruditorum* in 1682, journal where he published many of his results in calculus. Now, that is a good manner to avoid reviewers.

References

1. A. Csiszar, Troubled from the start. Nature **532**(306), 308 (2016)
2. E. de Schutter, Reviewing multi-disciplinary papers: a challenge in neuroscience? Neuroinformatics **6**, 253–255 (2008)

3. D.P. Peters, S.J. Ceci, Peer-review practices of psychological journals: the fate of published articles, submitted again. Behaviour. Brain Sci. **5**(2), 187–195 (1982). https://doi.org/10.1017/S0140525X00011183
4. D.F. Horrobin, The grants game. Nature **339**, 654 (1989). https://doi.org/10.1038/339654b0
5. J.M. Campanario, Rejecting and resisting nobel class discoveries: accounts by Nobel Laureates. Scientometrics **81**(2), 549–565 (2009). https://doi.org/10.1007/s11192-008-2141-5
6. J.L. Perez Velazquez, Scientific research and the human condition. Nature **421**, 13 (2003). https://doi.org/10.1038/421013a
7. J.T. Leek et al., Cooperation between referees and authors increases peer review accuracy. PLoS ONE **6**(11), e26895 (2011). https://doi.org/10.1371/journal.pone.0026895
8. D.F. Horrobin, The philosophical basis of peer review and the suppression of innovation. J. Am. Med. Assoc. **263**, 1438–1441 (1990)
9. D.F. Horrobin, Something rotten at the core of science? Trends Pharmacol. Sci. **22**, 51–52 (2001)
10. B. Brembs, K, Button, M. Munafo (2013) Deep impact: unintended consequences of journal rank. Front. Human Neurosci. 7:2091

Chapter 6
The Scientific Olympics: The Contest Among Scientists

In the course of doing science/research scholars fall prey to desires, they covet prizes, fame, fortune, which sooner or later leads to a most common human characteristic: rivalry, competition, perhaps envy. These features are fostered by the large number of scholars that populate the globe today — greater number than at any other past time— together with the limited funds available for research. The end result is that science lives in a state of hypercompetition.

6.1 A Question of Numbers

Mathematicians will never have enough time to read all the discoveries in Geometry, a quantity which is increasing from day to day and seems likely in this scientific age to develop to enormous proportions...

Christiaan Huygens, 1659

If Huygens was already complaining in those old times about the quantity of "discoveries", I am not sure what he would say today, when we are flooded with data, papers, journals, books, documentaries, videos, more papers... For sure he was right to forecast an expansion reaching "enormous proportions".

We are just too many academics these days. My reminiscences when I was a university student include my dialogues with the very few individuals who wished to pursue a Ph.D. degree regarding the laboratory or institution where we could go. Out of fifty or sixty, more or less, that finished with me the five year curriculum in Chemistry-Biochemistry at the University of Madrid, I would say those interested in continuing in academia were not more than 10 or 12, and perhaps not even 6 or 8 started the doctorate. Nowadays, I have seen what to me is an incredible number of students who want to become PhDs I say "PhDs" and not "academics" because, in truth and according to the experience in the institutes where I worked, quite a few of these students covet a degree because it will improve their chances to practise

© Springer Nature Switzerland AG 2019
J. L. Perez Velazquez, *The Rise of the Scientist-Bureaucrat*,
https://doi.org/10.1007/978-3-030-12326-0_6

medicine (that is, to be admitted into medical school), or being hired by corporations, and not because they plan to join the scholastic world. Nevertheless, whatever the case, too many PhDs are emerging for not too many academic posts available. It has been published that the number of doctorates awarded doubled between 1997 and 2017, at least in all the member countries of the OECD (Organisation for Economic Co-operation and Development). From 2003 to 2013, the percentage of science graduate students in the United States receiving a doctorate increased by almost 41%. However, these numbers were not matched with increasing academic positions. Only a fraction of these new PhDs will find a tenure-track job. As reported in "Education: the Ph.D. factory" [1]: "In 1973, 55% of US doctorates in the biological sciences secured tenure-track positions within six years of completing their Ph.Ds, and only 2% were in a postdoc or other untenured academic position. By 2006, only 15% were in tenured positions six years after graduating, with 18% untenured". The editorial "Spread your wings" [2], exemplifies the situation: "international science is training many more Ph.D. students than the academic system can support [...] Global figures are hard to come by, but only three or four in every hundred Ph.D. students in the United Kingdom will land a permanent staff position at a university. It's only a little better in the United States. Simply put, most Ph.D. students need to make plans for a life outside academic science. And more universities and Ph.D. supervisors must make this clear".

The article in The EuroScientist 'The "Lost Generation" of European Scientists: how can we make the system more sustainable?' talks about "the growing cohort of senior post-docs and other scientists who, after accumulating short-term contracts and temporary positions, find themselves excluded from the research system due to the lack of opportunities for permanent positions." In a meeting held at the ESOF (EuroScience Open Forum) 2018 Toulouse, the matter was discussed with great concern and questions were asked such as how to prepare young scientists for career shifts. The League of European Research Universities (LERU), an organisation with 23 members (as of 2018) that among other things attempts to enlighten politicians and policy makers about the situation and activities of research-intensive universities, has produced an executive summary of their Advice Paper "Tenure and tenure track at LERU universities: Models for attractive research careers in Europe" published in 2014. In this summary, they advocate for implementing the tenure-track system in Europe, something common in North America but very unusual in Europe —although some countries have a probation period but not quite the same as the tenure tack model. This system, they argue, allows for the introduction of formal autonomy —academic independence— at early stages of the research career and benefits from projectable career paths. Problem is that some governments control science and academia in almost a totalitarian fashion, in the sense that they offer the jobs in procedures like, for example in Spain, the call for 'oposiciones' (even though today there are a few private universities in Spain that do not depend exclusively in the 'oposiciones' system and can hire whoever they want); hence, continuing with the Spanish example, the only path for me to join a state university or the Consejo Superior de Investigaciones Científicas (CSIC) is to pass the 'oposiciones'. Therefore, if the tenure-track model is implemented, the national governments should grant universities the autonomy to experiment with it and learn from their experiments, and of course some financial support will be needed as well.

On account of today's technology, especially the internet, some are turning to freelance science. Of course, this is relatively feasible if one does not need a laboratory. But even if you do need experimental research, you could collaborate with a colleague who has the technology —I fully realise this is easier said than done. Doing research without the ties to an institution has the obvious advantage that managerial chores are avoided. It also allows one to fully concentrate on the theme that interests the individual most, embarking on projects that can be as long-term as one wishes as there are no preferred institutional research guidelines or milestones to achieve. Thanks to Internet and the explosion in communication technologies, collaborations among freelancers are possible. There exist social media systems like Slack, a social platform that allows people to create channels of discussions. Some disadvantages are obvious: lack of salaries and the potential feeling of loneliness are the two I would claim are most pressing for freelancers. Still, by joining a "virtual" organization like the Ronin Institute, which promotes independent scholarship, one can find weekly seminars, chats, collaborators, and it is even possible to apply for grants through the institute. Another initiative along the same ideas is CORES Science and Engineering that was formed in 2014. I transcribe the abstract for the Ronin Institute's presentation at the Performing the World Conference 2018 so that it is clearer what is meant by "independent scholarship": "The Ronin Institute (2012) is a self-organized community of scholars from the sciences and humanities formed with the idea that researchers should create their own measures of success. A non-profit organization, Ronin provides an affiliation for scholars, and a structure whereby researchers can apply for federal and state grants. This organization also cultivates organic scholarly cultures that focus less on career milestones and more on scholarly work and the conditions of the scholars doing that work. This develops new structures for scholarship rather than adapting to institutional expectations".

Competition in science is not new. There are myriad of examples throughout the history of science where one can witness the battle of egos. Some events are, should we say, very humorous, and their stories denote the expanded egos that many scientists had. We could start with the account of how Galileo's arrogance brought him into trouble, or the dispute between Newton and Leibniz over who had the priority in the invention of calculus, or Lord Kelvin refusing to the end of his life to accept J. C. Maxwell's theory of electromagnetism. Perhaps most humorous is the case of the Bernouillis, a family of mathematicians that ran from the middle of the 1600s to the early 1800s who had a vast influence in the development of mathematics; brother against brother (Jakob and Johann Bernouilli), and father against son (Johann and Daniel Bernouilli), are entertaining examples of hostility due to pride and arrogance: Johann, the father, published a book with all ideas on hydrodynamics of Daniel, his son, making sure he altered his book publication date to a few years before his son's publication finally appeared. In this appalling manner he stole ideas from his son. Thus one can see how obstinacy and vanity in science has flourished since times past, which resulted in some sort of competitiveness.

But in those old times the massive development of egos was somewhat curbed, although never completely stopped, for the attachment to the ego is a most rooted human feature. On the other hand, the vast infrastructure that today surrounds science and academia in general is a culture medium that promotes the growth of egos and the attention towards revenues and implementation of even more administrative procedures and infrastructure, attention points that were not the focus of the traditional academia, more concerned with research, discovery, and true scholarship, but, as it should be by now apparent after all said in this volume, those times are gone.

6.2   Corollaries of Academic Competition

There are several ramifications of the hypercompetition present today in academia. To start with one, competitive nature demands that grant proposals are rejected based on any possible reason, just like the grant that was rejected for using the wrong font (the story in https://www.nature.com/news/grant-application-rejected-over-choice-of-font-1.18686). Some numbers will illustrate the point: at the end of the 20th and start of the 21st century, the success rate for grant applicants to the Canadian Institutes of Health Research (the old Medical Research Council) was between 20 and 30%, and in some programmes could be as high as 50%, while today stays around 10%. In the U.S., the main body for biomedical research funding —the National Institutes of Health (NIH)— offers a similar trend, from a success rate of around 30% in the glorious past to the current percentage in the low teens, and some years even lower than 10%. "Statistical noise", as one of my American colleagues put it.

And, when we obtain a grant, often occurs that the money awarded does not increase with time. In 2003 I obtained an operating grant —that we used to perform experiments and cover some salaries in the laboratory— which gave us $35,720/year, and in the renovation of that project four years later, in 2007, I was awarded 32,435/year; furthermore, in the next renovation in 2013 they gave us $31,000/year. You see the dilemma, progressively less funds available concomitant with a progressive increase in prices and salaries for employees. How can we do research in this situation? The reason for the decrease in our awarded funds was made explicit by the funding agency, namely, a substantial increase in the number of applications that could not be matched by the corresponding growth in the agency's budget; as the Spanish aphorism goes: "de donde no hay, nada se puede sacar"

If to this scarcity we add the fact that organisations demand scholars to obtain a minimum number of grants or funds, then it is no surprise some scientists spend almost all their lives as PIs writing grants. Hence, young fellow, unless things change in the near future, be prepared to write anything between 5 and 10 grants a year upon entering the faculty in your institution. Of course to avoid this scenario one could be an eternal postdoctoral fellow, the problem is that there is a limitation in time to be a postdoc, beyond which time one has to change to, say, research associate, or perchance a technician; the natural course would be to become a PI, but you know what this entails… playing the grant lottery. The Nature News Feature

"Young, talented and fed-up" (https://www.nature.com/news/young-talented-and-fed-up-scientists-tell-their-stories-1.20872, *Nature* 538, 446–449, 2016) reveals the stories of a few early-career researchers in their path towards stress as independent investigators. Chief among the reasons for this stress is our current subject matter, the never ending competition of grant applications. And, sadly, obtaining funds for your research is not the end of your predicaments, for, as one of the young scientists put it in the aforesaid interviews: "It's stressful when you don't have money, and stressful when you do". The administration associated with the distribution and reporting of your funds was mentioned already in the first chapter of this volume.

Today's severe competition in science has had already a few damaging consequences for research. From scientific fraud to unfair and aggressive assessment of scholars' research outputs (recall the comments above in Sect. 3.2 on corporate culture in academia, where the cases of Stefan Grimm and the scam at the Karolinska Institute were mentioned) the realm of academia is becoming the playground of extreme contest. It is true that some fields are more prone to exhibit high competition among researchers. That is the case of molecular biology, as an illustration; if one is sequencing a genome, only the first to publish the sequence will carry the fame to posterity. Medical science, that is, research done in very practical approaches to treat diseases, is another field where high competition is present. It has been my experience that lay people tend to consider almost all biological research as being medical research, for whatever reasons, perhaps because the media are more focussed on widely reporting to the masses some extraordinary findings to cure this or that syndrome, or into the causes of such and such disease. But it should be stressed that medical science is a very especial field, and because of its lucrative aspects —just consider that pharmaceutical corporations invest vast amounts of funds with the hope to have something in return— and consequences in health care, it is here where one can find more cases of, say, misconduct or corruption. During my graduate training I had the opportunity to witness the ferocity of the competitive character. No need to go into details, suffice to say that very few years after I finished my degree, the department was almost completely renovated... because so many either left voluntarily or were forced to leave. I thus learnt, and since then I have tried to join places where competition is not that ferocious.

6.2.1 The Birth of the Scientist Salesman

Another corollary of the competition driven by the corporatization of academia is the rise of yet another type of scientist, perhaps an offspring of the scientist-bureaucrat: the scientist salesman. It should not be surprising. Finances, markets, require salespeople. Hence, the monetisation of academia brought about the birth of a researcher who, independently of how well he or she does research, is an expert at selling his/her research. And the tools are there: one can use those indices mentioned in Chap. 3 to bring attention to how the papers have ended up in very high impact factor journals, or one can announce the enormous funding the project has brought in, or the tremendous importance of the results for healthcare. Many of these announcements made by our salesman speaker giving a talk, for instance, will fall on not-too-experienced folks —those more versed in the topics of the talk may understand better when one is bragging and when not. If these individuals happen to be important administrators in charge of hiring personnel, they may be impelled to attract the speaker to their institute.

The media is another venue to sell your research. If the journalists do not use other peers to verify some of the allegedly ground-breaking results that our good scientist salesman is obtaining and how these will save the world, then the report may appear in newspapers and media in general, which may attract attention of more bureaucrats and politicians with little knowledge of science who may promote further these apparently marvellous scientific findings. The net result may be that our scientist is promoted, hired, or given a substantial amount of money that will cause any of the two previous happenings. In times of old there may have been too some exaggeration about one's research (for even da Vinci and Galileo needed to convince his benefactors to provide them salaries), but with today standards in research evaluation and related aspects that have been commented in this book, the scientist salesman has a great future. Of course, not all of us can do that. Not everybody can be a salesman, it is a matter of personality. But to some extent all scientist have to be, at least, a little bit of a salesman, if only due to the fact that we need to convince granting agencies to fund our proposals —recall the words in Sect. 5.2 on how we sort of lie when we write grant proposals. The sort of disappointing thing is that there is the need to, let's say, push the borders of veracity in several situations scientists face, like the abovementioned grant business or when one is looking for a job. Not that we have to lie indiscriminately, but we all have to play the games of our trade, as I have several times emphasised in this text as an advice to the newcomers.

In closing, this hypercompetitive climate is diverting attention from scholarship to finance and short-time thinking and research, and in addition to bureaucracy, it is limiting the potential that academics have in creating truly innovative research. We already made some comments in this regard in the section on corporate culture. Obviously, all these aspects, the advance of corporative features in academia, competition and shortage of funding, are inexorably interrelated.

Possible Solutions

If one major factor in the unsustainable hypercompetitiveness we face today is number of Ph.D.s, then the immediate conceivable solution is to limit the number of students entering graduate school. Institutions could widen the career paths for students interested in science, establishing information centres that advise young trainees. Along these lines, the NIH has a programme called BEST (Broadening Experiences in Scientific Training) which aims at preparing and informing trainees for a range of career options. In the course of my career I have seen a great number of graduate students who, in spite of striving to receive a Ph.D., were in reality not too keen to enter the ranks of academia, their real plans being in other spheres like medical school. There are quite a few other options where scientists, engineers, or scholars in general could have a prosperous career, although not in pure academia.

Especially in this age of Big Data, jobs are needed as data analysts, data scientists and data engineers, in fact one of my graduate students was hired as data scientist and has moved forward with success. My advice would be to identify at the beginning these students who desire degrees for reasons other than exploring and understanding natural phenomena, so that they do not need to spend 4 or 5 years working towards a doctorate —which many of them find laborious, tedious and many times agonising, as it has been my experience with this type of students, therefore in the end it would be good for them to avoid this waste of their lives pursuing something which they do not really enjoy— because in many of the alternative jobs these degrees like Master or Ph.D. are not needed. In this manner, the quantity of graduate students would start to diminish which would be an advantage for those really into science who crave for an academic position, as these could be better trained because of the smaller numbers: the lower the student/professor ratio the better the training.

A possibility to do some research in academic settings without having to fight for an academic post may be emerging in some readers' minds now, especially if you are a medicine student: work as a clinician but operate at the same time a laboratory. Now we come to the breed of scholars known as clinician-scientists. There are many, in fact. At least more popular in North America than in Europe, the clinician-scientist devotes the majority of the time to the clinic but has arranged with the clinical institution to have some time allotted for research... Or so they think. Because in the end the predicament of the clinician-scientist is the same as that of the today's scientist-bureaucrat: research is not really performed by them. As already explained in several sections, there is almost no time for the PI to do experiments, data analysis, proper thinking, that is, real research, and the solution is to apply for grants and obtain funds to hire trainees and personnel that will do the experiments for us, PIs. The situation may even be worse in the case of the clinician-scientist because there are also a bunch of administrative errands a clinician must execute. Hence, in the final analysis the research you would do as clinician-scientist is... how to say... research in your office about granting agencies and similar "investigations". You will be lucky if you can afford a weekly lab meeting where your personnel will inform you of the state of the lab. One also has to be certain about the commitment of the institution to the time allowed for "research"; I have known clinician-scientists whose research day (or days) were half spent in completing more clinical forms about the patients, and the little time left for research was, as aforesaid, research on how to obtain moneys.

Since a picture is worth many words, I reprint here, with permission from the authors, the figure on the scientific career pipeline that A. K. Lancaster and colleagues have published this year [3], where several alternatives to academia are shown. The authors elaborate on some possible solutions to "fix the pipeline", using their own words, and they touch on several points discussed in this book, like the rise of the big groups and the competition among scholars. To be precise, this is the currently existing pipeline that may not be the optimal, according to the authors (thus the need to fix it), because only those in the Academia part of the pipeline are considered curiosity-driven independent scientists, but in reality this should not depend on the type of job, that is, one may be a coordinator of research in a corporation and yet produce independent science, for instance, working at home developing a model or a theory. They conceptualise the scientific enterprise as being part of an ecosystem, which from sociological to scholarly aspects constitute the researcher's environment (for those interested, please see Fig. 2 in their paper); in this manner, they argue, the pursuit of independent science and the perception of the community as the scholar continuing to be contributing independent science is not dependent on your specific occupation.

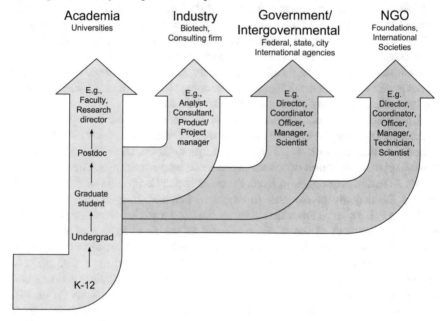

I have noticed how lay people react to the news of the relatively low number of research posts available and to the little money —compared with other trades— that is devoted to research, and the question that they many times have asked me is how come governments and policymakers do not invest much more in science, why is it there is so much money in, say, sports or movies, while only a trifle goes to basic science. This is the only real solution they see possible that can forever fix the impecunious researchers: make administrators put money in research. It is then when I have to bring their attention to the fact that basic research is not really a service, or at least is not an immediate service to humanity; we basic scientists do not provide a service in the short-term, although most likely all we do is of great service in the long, or very long, term. Society today asks for immediate rewards, short term services, very few can have the patience of an intellectual, the persistence to follow a route that may take years or decades in the investigation of some phenomenon. Basic research is not profit-oriented. The characters surrounding the impoverished researcher in Cartoon VIII (the "1-dollar cartoon") you saw above are all short-term service providers, the footballers entertain people and generate bets, the actors amuse audiences, the lawyers protect the accused as soon as needed, other characters make money go round... It is for these reasons they have immense amounts of capitals. My research to date, who knows, may be of service in the distant future or, perhaps, never. Hence, considering the contemporary state of the nations and the globe in general, who could be mad enough to give me as much money for my research as Juventus Football Club paid Real Madrid for Cristiano Ronaldo in the recent transfer.

To fully eliminate competitiveness is impossible because, first, it is imprinted in our genes —our biology dictates competition among individuals— and second, even if we could master the driving forces of our genes, the socio-economic situation is not about to change in the near future. If these two main determinants of human behaviour were to change, then it is not only in science but in any other realm of the endeavours of Man that competition may give way to, who knows, concord perhaps, serenity, equanimity. These are qualities with which scholars are supposed to be endowed, thus scientists could have an easier transition towards a peaceful world.

At least, one thing we scientists can do is to try not to cultivate and promote competition. It is harmful enough politicians promote our competition, but we are also fostering it; if you work in the field, you probably hear and read the word 'excellence' many, many times; it has become a very usual concept. I don't recall it was so popular two or three decades ago, but in these days all I hear is about maintaining excellence in research, excellence in teaching, in funding, academic excellence... It looks like we are all striving to be the excellent ones. And we create competitions too, awards for the best poster or presentation in a meeting or in a conference. It is reasonable to have a few awards, the important ones, like the Nobel or the Fields medal, things like these are fair, but when there is a myriad of small prizes given to students, postdocs and scholars for a variety of activities, well, it seems to me this promotes competitiveness, besides nurturing the egos, and, normally, where there are big egos, there are big problems. There was a point in the comments of some advanced medical students that I heard during a Buddhist meditation retreat where we were discussing the beauty of decreasing the self, one of them questioned whether the reduction of the selves was possible in their working environment, because as medical students doing residency they were constantly receiving compliments and awards, being told to endeavour to excel and become the best of all, nothing less. But now we digress towards a more complex and global issue, our contemporary (western) society being a culture medium for the selves, which could be the topic of many books.

References

1. D. Cyranoski et al., Education: the Ph.D. factory. Nature **472**, 276–279 (2011). https://doi.org/10.1038/472276a
2. Spread your wings, Nature **550**, 429 (2017)
3. A.K. Lancaster, A.E. Thessen, A. Virapongse (2018) A new paradigm for the scientific enterprise: nurturing the ecosystem. F1000Research 7:803. https://doi.org/10.12688/f1000research.15078.1

Chapter 7
The Future

The transformations occur precisely in times of crisis

Paulo Coelho, *O Vencedor Está Só*

All things appear and disappear because of the concurrence of causes and conditions.
Nothing ever exists entirely alone; everything is in relation to everything else

Buddha

Now in the twilight of my scientific career, when the original enthusiasm and energy have considerably diminished, I, more than ever, long for a place where I can devote myself without time pressures to the serene study of nature, leaving aside the haste for publication, the concern over funds or the excruciating bureaucracy. Many words in this text have been devoted to advice how to achieve that, how to navigate the system without becoming stuck in the bureaucratic closed loop. I think I have made it clear that it is my opinion that the inanities here revealed are crippling science today. It is difficult to escape the thought that science is turning into an industrial venture, as has been said several times here. Academia and the whole research system have been damaged but recovery is possible, the prescriptions are already being produced by eminent academics. I have accentuated in many sections above that we can all help with some initiatives. Scholars are voicing their discontent with politicians and bureaucrats, see for example the aforementioned open letter "They have chosen ignorance", a letter you can read and endorse by signing here in the EuroScience website https://openletter.euroscience. org/; it has 19322 signatories as of June 2018.

The main purpose of this chronicle has been to reveal the global perspective to the young student and lay person about where scientific research stands today, and not to provide a detailed account of how research in academia is organised… or disorganised. Other texts that provide similar views on these matters are Benjamin Ginsberg's "The Fall of the Faculty", or Sheila Slaughter and Gary Rhoades' "Academic Capitalism and the New Economy" where the fact that knowledge is today not a public good but rather a commodity to be capitalized on in profit-oriented activities is clearly exposed. It should be considered again that, as

© Springer Nature Switzerland AG 2019
J. L. Perez Velazquez, *The Rise of the Scientist-Bureaucrat*,
https://doi.org/10.1007/978-3-030-12326-0_7

was said in the Foreword too, my words have to be taken for what they are, a warning of what one will find in the execution of scientific research, and hopefully will not deter anybody from entering the ranks of science. Things are like they are, and one must adapt oneself to them. And to demonstrate that even in these times it is still possible to follow your interest and hunt for the answers you want to find, regardless of how very basic or pure the research may be, the Epilogue following this section presents the general strategy as to how I achieved it.

So, where do we go from here? It would not be surprising if, after reading this narrative, some felt that we are witnessing what some have called the end of science ("Career over curiosity: the end of science?" *Lab Times*, issue 3, 2016, http://www.labtimes.org/), or "The twilight of the scientific age" (M. López Corredoira, http://revistadefilosofia.com/54-09.pdf). Can this be true?

My opinion is that, while there have been extensive changes in the practice of scientific research, for nobody can deny this, some aspects of research have an immediate bright future, namely, the applied research. Nevertheless I do not think pure research will become totally extinct, although it is, in the present moment, in danger of extinction. The main reason for my optimism, if we can call it that way, is that the pair 'applied ∼ pure' research is a complementary pair —using the words of Kelso and Engstrøm (*The Complementary Nature*, MIT press)— where one needs/feeds on the other (it is clear the applied needs the basic science, but even the most basic research needs the applied, think about it in depth if you do not see it straightway).

In the same manner that there will always be artists, there will always be pure researchers, for, as aforementioned at the beginning of the text in the Foreword, artists and scientists —those who are not enslaved by utilitarian research and try to study natural phenomena regardless of the immediate practicality of their research— have important things in common, and among these is that the true artist and the true scientist cannot do any other thing but to follow their vocation.

In fact, this text has brought to attention reasons for optimism, that is, that pure and basic research will not evanesce, that scholars will continue pursuing their curiosity. There are many academics and young scientists who realise the situation and are trying to change it. It will not have escaped the grasp of the readers that vast number of scientists seem to be well aware of the problematic nature surrounding many scientific affairs, of the fact that policy makers have completely lost touch with the reality of research, of the unfair situation of ranking journals, institutions and individuals by quantifiers like impact factors and the like, of the hypercompetition. So, if almost all of us realise the present condition, why don't we change it? I am not sure why. Inertia comes to mind. One of my colleagues once used the term 'herd mentality'. This constitutes another paradox, perhaps the greatest of all exposed in this volume. Eliane Glaser, in her article "Bureaucracy: why won't scholars break their paper chains?" (https://www.timeshighereducation.com/features/bureaucracy-why-wont-scholars-break-their-paper-chains/2020256.article) exposes similar surprise as to why we don't change the system if almost nobody likes it, and advances some possible reasons like "form-filling provides relief from the messy challenges of research", or "[bureaucracy] offers a chimera of absolute transparency, consistency and fairness", reasons that may not be the real matter. She also points to the route to salvation, the way out of it: "Ultimately, resistance is impossible without collective solidarity". Precisely. It is in the collective realisation, agreement and behaviour where the only real possible solution stands. To some extent one has the impression that we behave as if we did not know, or decided to ignore, these things. It has been my experience that when I discuss these matters with some of my colleagues they voice their disapproval of impact factors or other elements surrounding research, and yet, a few days later they start telling me with great enthusiasm that they have a paper accepted in this or that journal of very high factor… What else can be said. Nevertheless, the initiatives talked about here, like the Bratislava Declaration, provide reasons to be optimistic. It will take time, for sure —we are dealing with monstrous bureaucracy— but I think some aspects will revert to what once academia was.

Different aspects have been touched upon in this volume. As a scientist, one would like to bind these facts together into some sort of coherence, finding a central theme. Conceivably, this central theme is the result of the current global (and national) economic situation, which together with the spread of administration and the deep involvement of politicians into the fabric of academia, is deviating scholarship from what once was. If one wants to change a situation, it is helpful to understand how it evolved. In general it is always a good strategy to study the origins of something to better comprehend the nature of the phenomenon. Therefore, understanding the chains of factors that caused the emergence of the present situation would help finding a general cure. So there is work here to be done by science historians and sociologists. I am neither, but if I may, I will venture to propose an overall course that led to this state. But my perspective, because of my ignorance, is relatively simple.

I think the current state was inevitable, it had to happen because science is immersed in society, and modern society is governed by financial concerns and bureaucracy. But wait, this has been occurring for centuries; perhaps today more accentuated, but nonetheless since the times of the Romans we can see how economy dictated the deeds of men. In times past, though, when few were doing research and the investigations were simpler, it was possible to avoid become submerged in the socio-economic flow (but remember that even the Galileos and Da Vincis needed benefactors, aristocratic or political sponsors who would pay them); you could retreat to your chambers and cogitate on scientific questions without much need for sophisticated laboratories. The time came when technology was available and the nature of the research became more complex, requiring heavy machinery, like cyclotrons, confocal microscopes… Money, and lots of it, thus became necessary. In parallel to the need for advanced laboratories, there was a large increase in the number of scientists, due in part to the massification of students in universities — more students need more teachers. These two ingredients foster competition, for funds or for fame, regardless, academia became more competitive. If to this we add the intrusion of bureaucracy and policy makers, then the emergence of the contemporary statu quo was imminent. Was there any resistance from scholars to this invasion, any battles fought? Not really. These administrative-minded individuals invaded academia because researchers could not say no, in fact at the start they were probably very welcome as these characters could pave the way towards acquiring more funds to buy more infrastructure and means for the increasingly complex experiments being carried out. Companies, industries, corporations may have been welcome in the beginning too, as they provided moneys for researchers; they became the new patrons of scientists, there was no need for aristocracy now. The ball started to roll, bureaucrats became more involved in planning research, companies invested more and more funds and naturally expected more in return, which in some areas like clinical research led to an unstoppable situation where hypercompetition is the norm, sometimes intertwined with misconduct. The rest is history.

If this, admittedly very simple, scenario was the origins of the new academic system, does it indicate a path or suggest a general solution that will fix the whole thing? The Possible Solutions subsections in each chapter have presented nothing more than specific counsels for particular obstacles a researcher may find in his/her career. As it was acknowledged in the Foreword, these advices are common knowledge to practising academics, but this text is for novices and lay audiences. Yet, those recommendations were all specific for various situations. One would like to find a global, general solution. As expressed in the previous paragraph, financial matters and its close relative corporatization are two main aspects contributing to the current state of affairs, but we have to live with them because these are fundamental aspects of modern society. Evidently, economy will not be remedied — that is a hopeless case, although it could be alleviated during very short spans of time— and politicians will continue to intrude, for governments have offices from where science is regulated and managed. Thus, the main hope seems to be to prevent further corporatization of academic institutions and to even start decreasing the number of administrative chores and administrators within academia. I am sure bureaucrats can do useful work in the science realm, but, please, limit your assistance to what scientists (not the politicians!) tell you they need. And even if we could dispose of moneys and bureaucrats, competition among scientists would still be present —albeit to a much less degree—because there are selves, egos, fame and fortune we all strive for. Therefore in my view, the best one can do is (as was recommended in several other parts of this text) to try to adapt oneself to the situation and avoid as much as possible the innumerable administrative chores academics have to perform these days, while at the same time using some workarounds to overcome specific obstacles. I will just end these comments saying that one thing that would help a lot the general situation in academia, in my opinion, is what was already mentioned in Sect. 3.1: finding a balance in many aspects, a fairer distribution of resources between big groups and individual researchers, between hypothesis-driven and question-driven projects, between utilitarian and holistic projects, a balanced, all encompassing evaluation of scholars considering work in the laboratory and administration, and more facets that, in my view, are today very imbalanced. Being a biophysicist I know that life exists because of nonequilibrium, but academia may benefit from a closer approach to equilibrium in these matters!

Improving the future normally goes through education, good education of children and adolescents should result in a better world, as education is the basis of society. In the specific case of research, the future will depend on who is taken to proceed towards Ph.D.s and postdoctoral fellowships. It is consequently essential to admit into the initial academic circles students who are curious about nature, creative, open minded and bold. These are normally the types of minds that transform how things are perceived and have the potential to revolutionise ideas, as George Bernard Shaw said: "The reasonable man adapts himself to the world; the unreasonable one persists in trying to adapt the world to himself. Therefore all progress depends on the unreasonable man." However, it is my opinion that the standard trends in the evaluation of students, particularly those at the graduate level, suffer from considerable myopia. While it is understandable that at lower levels of education (undergraduate), emphasis has to be placed on grades —since there is not much more on what to judge the young individual— I, and others, think that at more advanced education levels the relative relevance of grades obtained in courses should be diminished, and balanced with other measures of scholastic competence. The current educational trends favour what the polymath Heinz von Foerster called the "trivialization" of people (Understanding Understanding, Springer), that is, the creation of input-output individuals, where given an input, normally in the form of a question and after a supply of some information, a precise and determined output — an answer— is expected; such is the normally prosaic fashion that will guarantee an A in the exam. Thus, perfect grades could indicate perfect trivialization. It is my experience that students with very good grades do not make the best scientists, and by this term I mean those with a creative mind, mental faculties and genuine interest in finding answers to their questions. I have witnessed the extreme difficulties of truly remarkable students who had the misfortune of possessing enough creativity and mental autonomy to be non-trivial, reflected in not too notable grades, and thus were penalised accordingly for not being predictable citizens. This of course runs in parallel to the disregard of risky creativity by academic and funding troupes, many times mentioned in this volume. I thus advocate for a more balanced judgement of students, giving more weight to the references of mentors and to other measurements of achievements than to grades.

Then, finally, what would my advice be to those young fellows who are pondering to enter research and the world of academia? The first question I would recommend to answer clearly, with a yes or no, is whether you are passionate to discover reasons why natural phenomena occur in this way and not in another manner; in other words, if you want to live to understand nature or, on the contrary, if you want to have a job understanding nature in order to live. If you answer no to this first question —if you passion is not that great and if you prefer to work in order to live— then you could start reflecting about taking another route, not the academic one but perhaps the entrepreneurial, or the clinical, or, who knows, legal aspects like patent offices; the previous chapter presented these considerations about alternatives in science. But if you answered yes, if you cannot contain your fervour for comprehending nature, if you don't mind being underpaid while investigating natural phenomena, then I am afraid you have little choice, as little choice as I had when I finished my university studies, for I knew then that research was my life, my hobby. Then, my friend, you must do everything you can to enter academia, following the usual road of obtaining a Ph.D. degree, becoming a postdoc, and playing the rest of the games that we researchers, as much as any other people in other trades, have to play, games that have been described here. To what extent you want to play these games is up to you and your good understanding. The future cannot be accurately predicted, but some things can be sort of anticipated; keeping in mind the new developments that, as expounded in this text, are taking place in the minds of many scholars and institutions that are moving against the current in efforts to revert to true scholarship, an improvement in the scientific enterprise can be foreseen. It will take time, though. To a great extent, it depends on us, the scientists.

Epilogue

And if those last section's words of hope were not enough, let us end with one more encouraging note. A few paragraphs above I said that it is feasible to follow a more or less scholarly career, circumventing much of administration and games to be played. I may be an illustration of that, thus let me explain how things developed for me, and what I did and did not. Because, in spite of what the situation in science looks like in this time and age, you can still manage to devote yourself to your quest and avoid much of the cumbersome tribulations surrounding investigators. Where there is a will, there is a way. This was my way.

I first have to admit that I have been lucky enough in that, when I started in this business of research back in the 1980s, the tremendous corporative and administrative transformation of the scientific enterprise and academia had not reached the magnitude of today. There was still time to reflect, time to consider questions to ask of nature. I could still envisage a scientific life in which I would be able to research natural phenomena that attracted my interest and, more importantly, that I myself would be doing much of that research working at the bench, because this is what I like, I am an experimentalist and that's it.

Soon came one of my first encounters with the relatively new academic way of doing things: almost all that counts to be accepted to work in a laboratory and pursue a graduate degree is the grades one had as an undergraduate… plus some good relations, if one has them. Immense interest in research and curiosity about natural phenomena, such mine was, was not enough. The thought of going to industry came to my mind, but I already understood that research in industry is of another kind as the research I wanted to conduct: investigate my own questions and not those of the corporation that had lucrative end. Nevertheless, and after over 1 year of attempts at getting into a laboratory to complete a doctorate that would allow me to do independent research later on, I found a place, and came out a few years later with a Ph.D. It was during this time as a graduate student that I glimpsed at some basic features present in academia, or better yet, in academics. Thus I witnessed competition, rivalry, research strategies to reach specific answers about

© Springer Nature Switzerland AG 2019

J. L. Perez Velazquez, *The Rise of the Scientist-Bureaucrat*,
https://doi.org/10.1007/978-3-030-12326-0

natural phenomena —I learnt how to ask proper questions and elaborate a research plan— and the importance of funds. At the same time I was still seeing PIs doing experiments themselves and spending time in the laboratory, so no alarms yet were fired in my mind. Naturally, the distractions in those times were fewer too, as there was no internet, no emails, cell phones, constant communications, all these features so typical of today's communication era.

With this baggage I proceeded towards a postdoctoral fellowship, thanks to the courtesy of NATO, hoping to start investigating my own questions as soon as possible, which I did. I learnt here the importance to let trainees and other lab personnel develop their own approaches, ideas and interests, so it was the perfect laboratory for me. But it was at this time when bureaucracy and the new standards were fast infecting the scientific world. Realising these new patterns, and because I was too content being fully concentrated in my experiments, I thought I could delay the time to join the academic world officially as an independent scientist. Mind you, I was already one, because I developed many of my own projects and as well participated in others the laboratory required. This postponement of the inevitable — that is, if you want to be an independent scientist— proved to be a mistake. The reasons, which I could not have anticipated but I very soon realised when applying for jobs, were already mentioned in the Closed Loops Sect. 1.1: I had been a postdoc for too long and therefore my creativity and independence were suspected. I also discovered that the trends were not for generalists, rather institutions were looking for specialists in determined areas, hence my multidisciplinary formation —I was a mixture of biochemist, molecular biologist, biophysicist, electrophysiologist, and computational neuroscientist— helped little, if nothing (the current trend of high specialisation as a consequence of the new standards was mentioned in Sect. 3.2).

When I finally found an independent job —which took a while— I knew I had to start playing the games, described in this book, that would allow me to investigate my interests. Thus I started writing grants, publishing as many papers as I could, and filling up my CV with some of those free-of-charge jobs mentioned in Sect. 4.2. I also understood a common strategy of several of the newcomers to the academic world in order to improve their future: start looking for another job in a better (that is, more renowned) institution, shortly after being hired. But, in addition of not being too nomadic, I had a clear vision of what I wanted to do: research, doing experiments and data analysis and thinking. If my current institution allowed me to concentrate on these things, why should I move to a (more celebrated) place? I saw it coming, the lurking bureaucracy, so I withdrew as much as I could from panels, boards and other administrative duties, keeping these to a minimum; basically, I was managing my laboratory only, and this I did with great efficiency so that I could devote myself to doing experiments. The trick is to do the bureaucratic chores as fast and swiftly as possible, without becoming immersed in the aforementioned administrative loops. As well, one has to be careful with the new means to lose oneself: surfing the internet, attachment to email... It is so easy today to be diverted from true research! I experienced these distractions too but as soon as I realised it I altered my behaviour, curbing the craving of continuing probing the internet for more information; problem is, this control of behaviour has to be done on a very regular basis, hour after hour, day after day, such is the power of the communication business in our age. As well, little by little I found the grant lottery too tedious, so my grant applications declined over the years. I learnt too that utilitarian research is greatly favoured, as described in Sect. 3.1, so I had some typical projects that funding bodies liked because these could yield in the short-term some practical material. But there was always some unspent funds that could be allocated to the more adventurous projects we had in the laboratory.

The net result of the abovementioned attitude —the lack of participation in panels or in administration in general, the diminution of funds and other aspects I decided to cultivate in order to spend time in the laboratory— was not completely unexpected: my periodic appraisals were not improving, rather the opposite. Yet this did not bother me, because, again, I knew exactly what I wanted, and being the head of the institute was not one of those things (recall the advice in Possible Solutions in Chap. 1 of not climbing too high in the hierarchy). Of course these were my goals, I understand others who aspire to have more eminence will have to behave differently and play a bit more of the several games we have created, gaming which I reduced to its minimal expression. I also realised how I was penalised, in several occasions, for being precisely a full-time researcher. For example, one criticism to a grant I submitted was that I had included in a section of my CV more papers where I was first author than those as senior, or last, author, which was what the referees expected. It was explained in previous sections that authorship in papers is normally determined by the work done on the project, and because I am a full-time research worker I have done much of the work in several studies and therefore I was first author, but remember the rules of the game: I was supposed to be the senior PI and let other trainees work for me, hence I should have been almost always (at least since I became an established PI) the last author. I found too that the only publications that count are peer-review papers, so never mind it takes a long time and effort to write or edit a book, as I did in the past, these were seen as nothingness by my reviewers. Nonetheless, it was all fine for me, I was enjoying my research immensely. So you see, there is a way to avoid becoming the scientist-bureaucrat of Chap. 1.

But all good things come to an end, and, due to financial difficulties, my position was terminated. So I had to find a way to continue doing research and, at this time in my life now, almost completely suppress all the current preoccupations of the academic life. Thus I joined one of those institutions mentioned in Sect. 6.1 that foster independent scholarship. Now, finally, I have time to think, but I cannot do experiments because I do not have a laboratory anymore. I adapted to this new situation, thus I concentrate in my theoretical research that does not need benches and experimental subjects. I must admit that the discovery of time to do things carefully and gratifyingly has been one of my best findings, too bad I cannot publish a paper on that discovery. Another of what I consider among my very best findings is the finding of excellent scholars with whom I collaborated or exchanged views, and thanks to whom we pushed —to a very limited extent of course—the frontiers of knowledge. I had always admired those people of old that took years, sometimes decades, to complete a work, to achieve a goal. Today we are constantly with an eye on the clock, we are enslaved by the schedule, we have lost the capacity to be captivated by our task at hand. In science many of us are like pilgrims, slowly moving towards our goal; but always remember that in the pilgrimage, while the goal is important, the fundamental thing is to enjoy the road.

Printed in the United States
By Bookmasters